Biophysics of Mitochondria

Nikolai Vekshin

Biophysics of Mitochondria

Springer

Nikolai Vekshin
Moscow, Russia

Previously published with other publisher.

ISBN 978-3-030-33852-7 ISBN 978-3-030-33853-4 (eBook)
https://doi.org/10.1007/978-3-030-33853-4

This Springer imprint is published by the registered company Springer Nature Switzerland AG.
The registered company address is: Gewerbestrasse 11, 6330 Cham, Switzerland

Preface

Mitochondria are one of the most important organelles in aerobic cells. They have been studied during about 100 years by cytologists, biochemists, molecular biologists, and biophysicists.

The monograph summates a lot of important results on the investigation of mitochondria using a variety of biophysical methods: fluorescence spectroscopy, UV-vis spectrophotometry, IR spectroscopy, light scattering, optical microscopy, fluorescence microscopy, colorimetry, photobleaching, polarography, pH-metry, and so on. Especially, methodological issues are accented here.

A considerable part of the book is based on my own long-term experience. Also, I have presented main experimental data of many other researchers. I tried to analyze in depth alternative points of view, new hypothesis, and interesting contradictions.

I hope that this monograph will be useful for researchers and graduate students, specializing in mitochondrial biophysics, as well as in biochemistry, molecular biology, and cytology.

This book was published in Russian in 2019 by Photon Publishers (Pushchino, Moscow region).

I am very grateful to my colleagues and PhD students who participated in conducting the experiments with mitochondria over long time, namely, A.S. Lomtev, I.V. Sharova, A.N. Zinina, R.L. Anisimov, D.A. Shishmakov, I.B. Sokolova, V.I. Sukharev, V.I. Kovalev, E.A. Begunova, D.N. Kurdyukov, K.E. Krekhova, M.A. Frolova, A.V. Chaplygina, A.M. Lvov, and many others, listed as coauthors in joint publications.

Moscow, Russia
September 2019

Nikolai Vekshin

Contents

Abbreviations

MC	mitochondria
PMC	proto-mitochondria
PostMC	post-mitochondria
LF	lipofuscin
NAD	nicotinamide adenine dinucleotide
NADH	nicotinamide adenine dinucleotide reduced
NADPH	nicotinamide adenine dinucleotide phosphate reduced
FMN	flavin mononucleotide
FAD	flavin adenine dinucleotide
RF	riboflavin
ATP	adenosine triphosphate
ADP	adenosine diphosphate
AMP	adenosine monophosphate
EDTA	ethylenediaminetetraacetic acid
SDS	sodium dodecyl sulfate
RS and RI	rotenone-sensitive and rotenone-insensitive
DCPIP	dichlorphenolindophenol
pNTV	para-nitrotetrazolium violet
PA	palmitic acid
DNA	deoxyribonucleic acid
TPP	tetraphenylphosphonium
TMR	tetramethylrhodamine
UV	ultraviolet
DTT	dithiothreitol
PAAG	polyacrylamide gel
MDA	malonil dialdehyde
TBA	thiobarbituric acid
BSA	bovine serum albumin

ANS	aniline-naphthalene sulfonate
M540	merocyanin-540
7AAMD	7-amino-actinomycin D
P	polarization degree
τ	lifetime of the excited state

Chapter 1
Introduction

1.1 Myths and Reality

Over the years the work with mitochondria (MCs) in vitro, in situ and in vivo, lead the author to develop his own point of view, at some points markedly different from what is currently generally accepted. Below, the author has been formulated some of modern views (in italics in quotation marks; the examples are in [1, 2]) and give an alternative author's interpretation [with references to own publications].

1. *"MCs have the form of shoes or worms."* MCs in cells of animal organs (liver, muscles, kidneys, heart, brain, etc.) are spheres or spheroids, the same as in an isolated form [3–5]. "Shoes", observed by electron microscopy, are often the result of stitching two or three MCs during processing and fixation of a sample. "Worms" in optical microscopy is fibrils, on which MCs sit, or a consequence of the uneven distribution of used dye.
2. *"MC membranes are lipid bilayer in which proteins float."* Since the protein-lipid ratio in the inner membrane is 3:1, giant protein-lipid complexes with small lipid impregnations really exist [6].
3. *"The size of MC equals to several microns."* No, it is usually 1 ± 0.5 micron [3–5]. Also, small germinal proto-MCs (PMCs) with a diameter of 0.2–0.4 micron are present in cells [3].
4. *"There is a MC network, a giant branched reticulum in a cell".* The MC-reticulum is absent [3–5]. In electron microscopy, "giant" MCs are the result of cross-linking of a number of MCs by formaldehyde, glutaraldehyde, etc. In fluorescence microscopy with dyes, they appear due to staining of fibrils or actin filaments on which the MCs sit [7, 8].
5. *"MCs arise by division."* It occurs usually not by division. DNA molecules from old MCs enter into cytoplasm and complete the membrane around themselves (in cooperation with nuclear DNA), resulting in appearance of germinal PMCs, less than 0.2 microns in size [3, 9]. That is why, unlike MCs, lipofuscin – aging pigment – was not found in PMCs of old animals [22].

N. Vekshin, *Biophysics of Mitochondria*,
https://doi.org/10.1007/978-3-030-33853-4_1

6. *"The main function of MCs is the synthesis of adenosine triphosphate (ATP)".* This is only one of many important functions, among which are: metabolism of di - and tricarboxylic acids, heat regulation, calcium deposition, electrical activity, etc. [1, 2]. Moreover, various functions dominate in different organs [1, 3], for example, in adipose tissue – almost only one heat production, and in neurons – generation of electric potential.
7. *"The energy of electron transport through respiratory chain is needed for synthesis of ATP."* No any special big energy is needed for the act of phosphorylation of ADP to ATP [10]. ATPase, like all enzymes, works itself in both directions. A little energy is needed only for ATP desorption.
8. *"Electron is transmitted sequentially along a long respiratory chain."* The chain is not sequential and is not always long, but usually 2–3 short parallel chains exist [19].
9. *"Oxygen consumption takes place on cytochrom oxidase."* Not only on cytochrome oxidase. Normally, during the oxidation of nicotinamide adenine dinucleotide reduced (NADH) by isolated MCs, about of 0.1–5% oxygen can be consumed on NADH oxidase, giving superoxide [1]. Up to 40% of oxygen can be consumed on NADH oxidase, if the respiratory chain is damaged [19].
10. *"In the outer membrane, there is rotenone-insensitive NADH dehydrogenase, transferring electrons through cytochrome C to respiratory chain of the inner membrane."* There is no such enzyme [11]. The reaction of rotenon-insensitive oxidation of NADH is carried out by fragments of NADH dehydrogenase of the internal membrane, and this occurs in damaged MCs.
11. *"All changes in absorption and fluorescence of NADH and flavins of MC occur due to oxidation of NADH and reduction of flavins."* Not always. For example, upon binding to dehydrogenases, the quantum yield of NADH fluorescence increases, which creates the illusion of restoring NAD to NADH [7]. After the exit of flavin from NADH dehydrogenase, its absorption (at 450 nm) and fluorescence (at 520 nm) increase sharply [12, 13] due to elimination of contributions of light reflections on membranes [5] and Tindal hypochromism [14]. In "classical" works by B. Chance and others, this was not taken into account.
12. *"MC has two membranes – external and internal".* This is true for MC in a cell. But isolated MCs quickly lose their outer membrane [5] if there is no EDTA or succinate in the medium. At room temperature, the outer membrane breaks and flies (in the absence of EDTA or succinate) during 10–15 min. In a living cell, germinal PMCs with a diameter of less than 0.2 micron have only one membrane.
13. *"Restoration by substrates (NADH, succinate, etc.) of artificial electron acceptors takes place with participation of flavins, iron-sulfur clusters and ubiquinone."* It is hardly possible that flavins or iron-sulphur clusters, or ubiquinone, are participate [11, 15]. Restoration such acceptors as ferricyanide, dichlorophenolindophenol (DCPIP), tetrazolium etc. by substrates is a high-speed shunts.
14. *"Old MCs in a cell are digested by lysosomes".* This is only one of the ways of degradation of old MCs, and not the main. In specialized animal cells (brain, muscles, heart, liver), MCs degrades to lipofuscin granules [3, 16], which are not

utilized, but remain in the form of intracellular "garbage". In some organs (kidneys, lungs) old MCs degraded to post-MCs – hollow spheres with a diameter of 2–6 microns, which are lysed by protease or out of cells [3].

15. *"MC can be specifically stained with mitotrackers or rhodamine 123"*. Mitotracker and rhodamine 123 have no high selectivity [3, 7, 8]. They can stain not only MCs, but also actin filaments that create the illusion of a worm-like MCs or a giant network. These dyes are poorly penetrating into the intact MCs, but good paint the damaged MCs [17]. Among rhodamine dyes, Janus green has the greatest selectivity to MCs. Real markers of MCs are: succinate dehydrogenase which reveals reduction of tetrazolium to formazan, deposited inside the MC [3], and DNA, stained with ethidium bromide.
16. *"Exposure of MCs on light causes they damage"*. This is true for the intense UV radiation, leading to damage of membranes and formation of photo-lipofuscin [16]. Non-intensive light irradiation activates the electron transport, oxygen consumption and ATP synthesis [18].
17. *"Fuorescent probes can easily measure the transmembrane potential."* Not easy. Although positively charged dyes bind with MCs better than negatively charged ones, it is not a matter directly to transmembrane potential, but can be a result of the presence of negatively charged phospholipids and proteins on the outer surface. Damaged MC bind much more dye molecules than non-damaged [17].
18. *"There are special carriers for substrates in the inner membrane: succinate-translocase, ADP-translocase, NADH-translocase, etc."* Isolated MCs have high permeability, for example, for succinate that cannot be explained by any carrier [19]. In the hypotonic environment, permeability increases strongly, for example, for NADH. Transport of substrates into the MC occurs simply across the concentration gradient due to the presence of large pores in the outer membrane, and in the inner – small nonspecific channels. This process is no different from transport of artificial acceptors and dyes, which certainly no special conveyors.
19. *"Three or four free protons are taken from matrix to create the transmembrane potential."* There are no free protons in matrix. If we calculate the number of protons inside the sphere of 1 micron, by the same way as Lehninger did [24], we get ~0. Protons are not taken from matrix, but from the substrate, as correctly stated in the original Mitchell's scheme [20].
20. *"Flavin and iron-sulfur clusters are obligatory necessary for NADH oxidation in the first complex of respiratory chain"*. Many dehydrogenases (alcohol dehydrogenase, etc.) have neither flavins nor iron-sulfur clusters, but oxidize the NADH molecules and even restore ubiquinone [23]. When placing the MCs in water, the flavin molecules fall out of them [7, 12], but NADH is perfectly oxidized, and this is partially blocked by rotenone [19].
21. *"Once upon a time, at the start of evolution, aerobic bacteria infiltrated a proto-cell and turned into MC."* Not the fact. The DNA of aerobic bacteria weighs millions of Daltons, and the DNA in MC is a thousand-fold smaller. It is more likely that it was not an introduced bacterium, but a DNA-containing virus or phage. The above material was partially published [21].

1.2 Mitochondria in Cell

In aerobic cells of eukaryotes (Fig. 1.1) a special place is occupied by MCs – small (about 1 micron) organelles performing a lot of various functions, among which one of the most important is the implementation of aerobic respiration – oxidation of substrates (NADH, di – and tricarboxylic acids) by molecular oxygen [25–27].

Due to the oxidation, an electric transmembrane potential and ATP are formed, which are used as energy sources for intracellular reactions.

All autotrophs (photosynthetic plants) and heterotrophs (fungi, animals) have MCs.

The number of MCs in different cells varies greatly: from a dozen pieces (in yeast) to several thousand (in neurons). One cell of the rat liver contains 500–800 pieces. All cells of such organs as the brain, heart and muscles (having intensive aerobic metabolism) contain several thousand MCs, occupying up to 20% of the intracellular volume [25–29]. Oocytes, having a very large size, contain up to 300,000 MCs. Large amoeba *Chaos chaos* has 500 pieces. In intestinal anaerobic ent-amebas and parasitic protozoa, characterized by anaerobic metabolism, MCs are absent or only a few pieces.

In specialized cells of plants and animals, three types of mitochondrial organelles exist simultaneously (in comparable amounts): young small (0.1–0.45 micron) proto-MCs (PMCs), mature MCs (about of 1 micron) and the old post-MCs (1–6 microns) [30].

MCs are able to pass within a cell to the areas of greatest energy consumption [27–31] moving through the cytoplasm due to diffusion or using cytoskeleton.

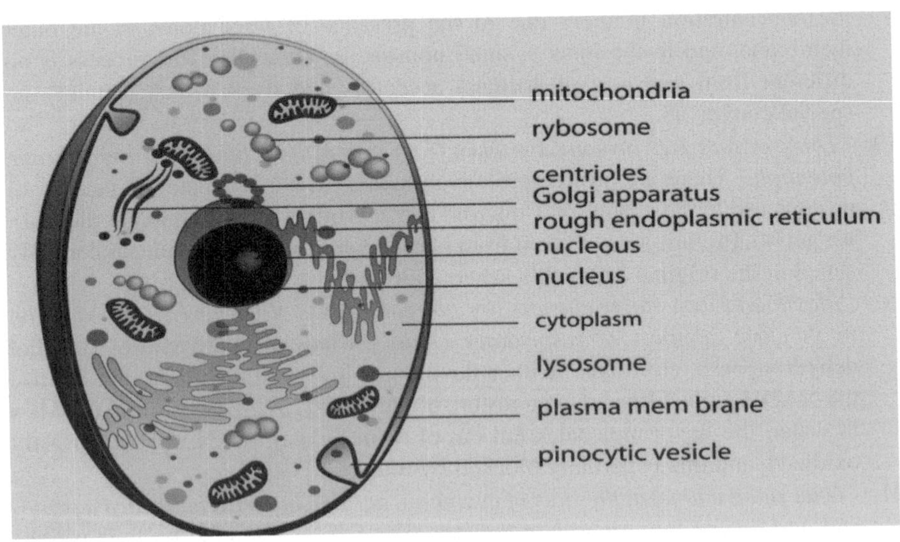

Fig. 1.1 The animal cell and its organelles

Moreover, sometimes they can be transferred from one cell to another, for example, from astrocytes to neurons [32].

In order to be sure that the observed structures are indeed MCs, it is necessary to make sure that they have a negative charge (other intracellular organelles are positively charged or not-charged), succinate dehydrogenase (only the respiratory chain contains this enzyme), cytochrome oxidase (this copper-containing enzyme is characteristic of respiratory chain), cardiolipin (20% of phospholipids; it almost absent in other organelles) and circular DNA ~16 kilo-bases.

1.3 Early Study

The word "mitochondria" originates from two Greek words: μίτος – mitos – thread and χονδρος – chondrios – grain. The duality and inconsistency of this term are associated with the fact that some researchers attributed the function of intracellular respiration to thread-like structures, and others – to spherical. Some called them threads or worm-like bodies, and others – chondriosome, chondrosphere, etc. To reach consensus, they began to use dualistic terms – filamentary grain, plastopedia and finally, mitochondria [33].

In 1850, histologist Kellicker observed MCs in the form of granules in a muscle cell. He was the first who isolated MCs (from insect muscle) and showed that they have a membrane and swell in water. Other researcher Retzius called these granules sarcosomes.

In 1890, Altman [34] proposed a specific method of staining mitochondrial granules, which he called bioblasts. He wrote the book "Elementary organisms", in which he suggested that bioblasts are autonomous units, and he postulated that they are "elementary living particles".

Many years later, Lehninger jealously described his book as "fantastic and highly speculative," although many researchers, including Lehninger himself, borrowed from it a number of ideas and methods.

Altman believed that bioblasts, like bacteria, are capable of independent existence and growth in the form of colonies inside the cell. He also expressed two very strong assumptions: about genetic role of these self-reproducing granules and their active participation in metabolism [34] (the both hypotheses were subsequently fully confirmed).

In 1898, Benda introduced the term *mitochondria* and used crystal violet for coloring. Michaelis used for coloring Janus green. Michaelis proved that MCs play a major role in intracellular respiration. He also showed that MCs in living cell can cause changes in the color of dyes due to red-ox processes.

In 1908, Rego, who studied reactions of MCs with dyes, concluded that these organelles contain phospholipids and proteins. His method allowed to reveal special chemical properties of MCs in cells of various types and to differentiate them from other cytoplasmic granules.

In 1912, Kingsbury suggested a possible role of MCs as intracellular oxidation centers. He considered them as structures for metabolism of substances in cellular respiration.

In 1913, studying the oxygen consumption in liver cells, Warburg found sub-cellular particles, which he called granules (it were MCs). He suggested that oxidizing enzymes for reactions in which the final products of glucose cleavage are further oxidized to carbon dioxide and water are associated with these granules.

In an attempt to identify the biochemical changes that occur during transformation of normal cells (with controlled growth) into cancer cells (with uncontrolled growth) Warburg measured the rate of oxygen consumption in tissue sections. He found that cancer cells produce abnormally large amounts of lactate, and concluded that they use the anaerobic pathway of glucose metabolism. He suggested that normal cells are transformed into malignant cells due to lack of oxygen. Warburg also observed that aerobic respiration is inhibited by poisons such as cyanide. He believed that such substances can cause cancer.

At the late of 20-ies, Warburg discovered the respiratory enzyme cytochrome oxidase, catalyzing oxidation reactions in MCs. Using a homemade spectrophotometer, he found that the active coenzyme of cytochrome oxidase is porphyrin with an iron atom (and copper), acting as a carrier of electrons to oxygen.

The aerobic pathway of ATP synthesis in MCs was discovered in 1931 by Engelhardt and called oxidative phosphorylation (the oxidation energy is "stored" in the terminal pyrophosphate bond of ATP).

In 1939, Belitzer and Tsybakova showed that ATP synthesis is associated with the transfer of electrons through respiratory chain.

In 1948, Kinday and Schneider found a complete set of enzymes of the Krebs cycle in MCs.

In 1949, Lehninger showed that oxidative phosphorylation is carried out exclusively in MCs.

In the 50s, Green and Richterich found in MCs a number of enzymes for the oxidation of proteins, fats and carbohydrates to the substrates of the Krebs cycle.

Soon after, Chance and Williams showed that NADH dehydrogenase, succinate dehydrogenase and cytochrome oxidase, exist only in MCs, and strongly associated with structural proteins.

In 1965–1968, a number of authors showed the energy-dependent release of protons from MC (in support to chemiosmotic hypothesis of Mitchell) [35, 36].

In 1967–1969, Levy et al. and Schneitman et al., as well as Hoppel and Cooper using a "soft" detergent digitonin selectively destroyed the outer membrane of MC. As the result, it has been possible to divide the external and internal membranes, and the content inter-membrane space and matrix.

References

References for Section 1.1

1. Cox M, Nelson D (2017) Principles of Lehninger's biochemistry. Moscow
2. Chentsov YS (2004) Introduction to cellular biology. Akadem, Moscow, pp 324–355
3. Belyakovich AG (1990) Study of mitochondria and bacteria using the tetrazolium salt pNTV. ONTI, Pushchino
4. Zinina AN (2008) A brief history of ideas about the structure and functions of mitochondria. Photon, Pushchino
5. Kurdyukov DN, Vekshin NL (2016) Biophysics 61(4):736–743
6. Vekshin NL (1988) Biophysics 33(2):360–362
7. Vekshin NL (2008) Fluorescence spectroscopy of biopolymers. Photon, Pushchino
8. Vekshin NL, Frolova MS (2014) Biophysics 59(5):1034–1039
9. Begunova EA, Vekshin NL (2015) Biophysics 60(6):1109–1117
10. Vekshin NL (1990) Com Mol Cell Bioph 7(1):17–25
11. Sharova IV, Vekshin NL (2004) Biophysics 49(5):814–821
12. Sokolova IB, Vekshin NL (2008) Biophysics 53(1):73–77
13. Frolova MS, Vekshin NL (2014) J Fluoresc 24:1061–1066
14. Vekshin NL, Frolova MS, Kovalev VI, Begunova EA (2015) Biophysics 60(1):129–135
15. Lomtev AS, Sharova IV, Dokl VNL (2001) Acad Sci 376(1):114–116
16. Frolova MS, Surin AM, Braslavsky AV, Vekshin NL (2015) Biophysics 60(5):1125–1131
17. Vekshin NL (2013) Biophysics 58(6):1074–1080
18. Vekshin NL (2002) Photonics of biopolymers. Springer
19. Lvov AM, Vekshin NL (2019) Biophysics (in press)
20. Skulachev VP (1985) Stories on bioenergetics. Moscow
21. Vekshin NL (2017) Biophysics 62(6):1232–1234
22. Krehova KE, Vekshin NL (2018) Biophysics 63(2):265–269
23. Kamensky KA, Vekshin NL (2018) Prikl Biochem Microbiol 54(3):316–319
24. Lehninger A (1974) Biochemistry. Mir, Moscow

References for Section 1.2

25. Lehninger A (1966) Mitochondria. Mir, Moscow
26. Furalev VA (1998) Cytology: structure and function of cell organelles. OL VSMS, Moscow
27. Cox M, Nelson D (2017) Principles of Lehninger's biochemistry, vol 2. Moscow
28. Ozernyuk ND (1978) Growth and reproduction of mitochondria. Nauka, Moscow
29. Luzikov VN (1980) Regulation of mitochondrial formation. Nauka, Moscow
30. Belyakovich AG (1990) Study of mitochondria and bacteria using the tetrazolium salt pNTV. ONTI, Pushchino
31. Zaguskin SL et al (2007) Cytology 49(10):832–838
32. Hayakawa K et al (2017) Nature 535(7613):551–555

References for Section 1.3

33. Lehninger A (1966) Mitochondria. Mir, Moscow
34. Altmann R (1890) Die Elementarorganismen u. Beziehungen, Zellen, Lpz
35. Zinina AN (2000) Brief history of development of understanding of structure and functions of mitochondria. Pushchino
36. Cox M, Nelson D (2017) Principles of Lehninger's biochemistry

Chapter 2
Main Properties of Mitochondria

2.1 Structural Features

Two forms of MCs are most common in cells [1–3]: spherical (most often) and ellipsoid. Historically, the MCs on schema are depicted usually in the form of ellipsoid "shoes" (Fig. 2.1).

The MC matrix is surrounded by two membranes. The outer has a thickness of about 70 Å. Its main function is separation of MC from cytoplasm. In this membrane, there are many channel-forming proteins – porins, forming pores with a diameter of 20–30 Å, through which small molecules can easily penetrate (up to 5 kDa). This membrane is characterized by the presence of mono-oxygenase, acyl-CoA synthetase and phospholipase A2.

The outer membrane is separated from the inner at the distance of 100–200 Å. Where the membranes touch, a protein receptor places, which facilitates the transport of proteins encoded by nuclear DNA into matrix. One of the proteins, freely floating in the inter-membrane space, is cytochrome-C. Inner membrane forms numerous comb-like folds – crystaes (from lat. crista – comb, Fig. 2.2), significantly increasing the surface area to accommodate multi-enzyme complexes.

Sometimes mitochondrial membranes are depicted in schemes in the form of lipid bilayers, in which proteins "float". Such models are poorly consistent with reality, because the outer membrane protein/lipid ratio equals to 2:1, and in the inner one – 3:1. If we take 3:1 and average geometric sizes of phospholipids and proteins, the inner membrane on the mental section looks like a set of protein globules, and lipids are only interspersed between them [4]. In reality, proteins are not evenly spaced, but form complexes with each other.

A characteristic feature of the composition of inner membrane is the presence of cardiolipin [1, 3] – a special phospholipid containing four fatty acids and making the membrane poorly permeable to protons and cytoplasmic substances even of low molecular weight. The inner membrane is very little permeable for NADH. It is well permeable for succinate and oxygen.

N. Vekshin, *Biophysics of Mitochondria*,
https://doi.org/10.1007/978-3-030-33853-4_2

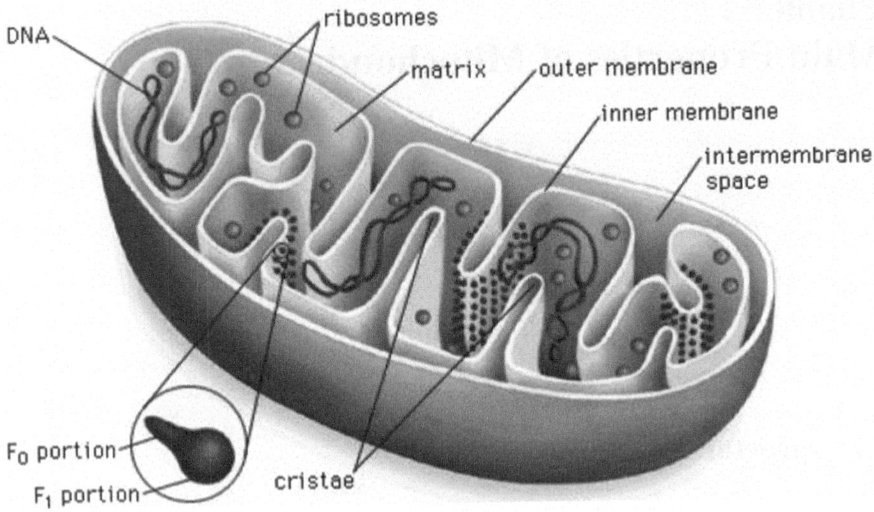

Fig. 2.1 Scheme of the internal structure of mitochondria

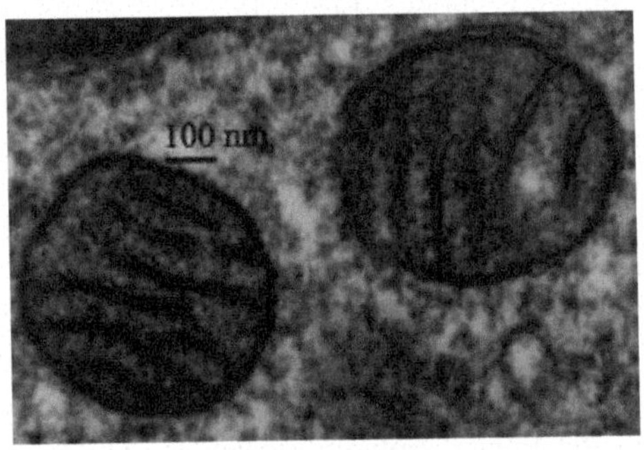

Fig. 2.2 Electron microscopy of two human mitochondria in cross section. Cristae inside matrix are visible

The inner membrane, in addition to structural proteins, contains a number of proteins of the respiratory chain and proton-sensitive ATPase, synthesizing ATP at the matrix side.

Inside the MC, a "jelly-like" matrix contains the enzyme oxidation systems of pyruvate and fatty acids, as well as enzymes of tricarboxylic acid cycle (Krebs cycle). Matrix contains also ribosomes, mitochondrial DNA and RNA.

2.2 Giant Mitochondria and Reticulum?

In several publications, e.g. [5, 6], it was reported about giant MCs in cells of liver, heart, etc. These "giants" permeating the entire of cell, were obtained by layer-by-layer reconstruction of processing sections with electron microscopy.

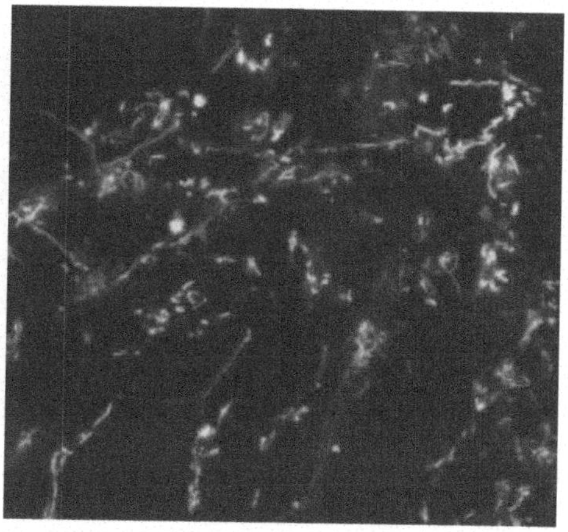

Fig. 2.3 Nuroblastoma cell stained with dye Mito-tracker Green. Confocal microscope Leica – TCS SP5, 405-nm excitation, green fluorescence (Frolova M. S., Vekshin N. L., 2018)

Giant MCs were "opened" in the 80s, but were almost universally forgotten in the 90s. The fact is that cells of the heart, muscles and other organs are literally stuffed with thousands of MCs, localized close to each other, and for fixing tissue sections in electron microscopy were used glutar-aldehyde, formaldehyde (formalin), etc., which, as it turned out, chemically sewed closely spaced organelles. . .

Although in some cases giant MCs can probably occur, for example, in pathologies, but in normal cells of animal organs they are usually absent.

Mitochondrion – so called a "giant" MC in the sperm cell. The size of this mitochondrion in reality is only 3–4 micron in length and 1 micron in width, so this MC would be more correct to call large, not giant.

With the help of optical microscopy and fluorescent dyes, some researchers "discovered" in cells a mitochondrial reticulum – giant network.

For example, in [7] using rhodamine 123, a set of thin long (up to 50 micron) curved filaments, interpreted as filamentous MCs, were revealed inside the cells. Although, due to existence of the positive charge, rhodamine 123 indeed able bind to negatively charged MCs, but, unfortunately, not only with MCs. This dye is capable of very intense staining of fibrils [8], on which MCs sit. Moreover, the suspension of isolated MCs was poorly stained by this dye [9].

Let see another example. Dye mito-tracker has (by its very name declares) a high specificity for MCs. But it's not just MCs. For example, when painting of neuroblastoma cells with green mito-tracker detected (Fig. 2.3) two types of structures: spherical (this is MCs) and filamentous (this is fibrillar filaments; they have neither succinate dehydrogenase nor DNA) [10].

Figure 2.4 shows photographs of the neuroblastoma cells, obtained in the "green fluorescence field" (it belongs to flavoproteins of MCs) without any dyes, and in the "red fluorescence field" while painting by 7-aminoactinomycin D (7AAMД is probe for DNA). In both cases, only spherical MCs (about 1 micron in size) are visible. No filamentous structures are detected. Both pictures are perfectly superimposed on each other.

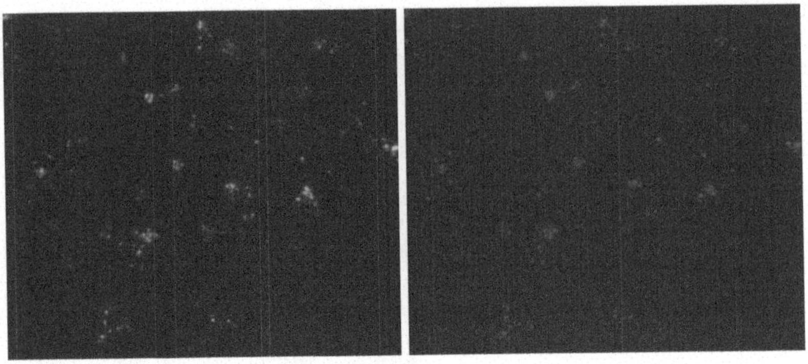

Fig. 2.4 Flavin's green fluorescence (left) of mitochondria of neuroblastoma cells and red fluorescence (right) of the cells, staining with 7AAMД (Frolova M. S., Vekshin N. L. 2018)

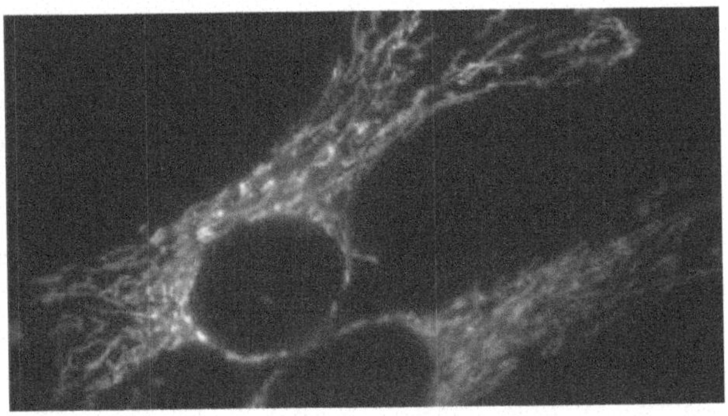

Fig. 2.5 Two *HeLa* cells, painted with "mito-tracker" mtGFP (Nano Lett., 2008, 8 (11), 3887–3892)

The textbook [11] at page 327 shows a photo of "mitochondrial reticulum", identified by fluorescent rhodamine. In fact, it is painted fibrillar threads.

Figure 2.5 shows "mitochondrial reticulum" in two HeLa cells stained with mtGFP. Bright dots of size 1 micron it's probably MCs, and the threadlike structures are not MCs.

In order to verify whether the studied intracellular structures are mitochondrial or not, it is sufficient to conduct the succinate: tetrazolium-reductase reaction (succinate dehydrogenase exists only in MCs) [9] or to color the cell with ethidium bromide, or Hoechst, or 7AAMD, which intensively fluoresce only upon binding to DNA, but not with proteins [12]. Microscopists who did not do such experiments, for example [7, 11, 13, 14], came to the wrong conclusion about the presence of filamentous MCs.

Although MCs are plastic organelles and, generally speaking, are able to change own shape, but in specialized animal cells (liver, brain, muscles, heart, kidneys) they prefer to be spheres with a diameter about of 1 micron. When researchers in the application of some dyes see a variety of non-spherical particles of about 1 micron, this is often a consequence of the uneven distribution of the dye, as well as the internal heterogeneity of each MC.

Many MCs sit on fibrils, but are not a reticular network in themselve. Reticulum exists only in the functional sense: the outer membranes of many MCs, connected by fibrils, can, in principle, be a single electrical network [15].

Aging MCs lose their charge [16, 17], so in the case of the presence in the cell of many hundreds old MCs (post-MCs) they can aggregate, creating the illusion of a giant MC.

2.3 Evolutionary Origin

It has long been observed that MCs are very similar to bacteria. The monograph [18] lists the main similar features. In particular, MCs and gram-negative bacteria have two membranes. The composition of membrane lipids in those and others is very similar. Their respiratory ensembles are localized in membranes. Ribosomes have similar sedimentation coefficients. The sequence of nucleotides in ribosomal RNA of MCs is more similar to that in bacteria than with r-RNA of cytoplasm of the same cells. Their DNA, unlike nuclear DNA, has a ring structure and is replicated semi-conservatively.

It is important not only the similarity of MCs with bacteria (size, shape, the presence of DNA, own ribosomes, respiratory chain, etc.), but also that the MC genome is not self-sufficient: it encodes only some subunits of own proteins, while others subunits are encoded by nuclear DNA, synthesized in cytoplasm and then transported to MC.

According to the theory of symbiogenesis, MC evolved as a result of the capture of some aerobic bacteria by primitive single cells. This allowed the cells to use oxygen to generate energy and accelerate their metabolism. It is assumed that the invading bacteria transferred many of their genes into nuclear DNA or even formed nuclear DNA, making eukaryotic prokaryotes [19, 20].

The hypothesis about the origin of MCs from bacteria-endosymbionts was first expressed with the same perspicacious by Altman (1890). Its essence is as follows: in the Proterozoic era with the advent of photosynthetic bacteria in the earth's atmosphere began to accumulate oxygen – a byproduct of their metabolism. A part of anaerobic heterotrophs moved from oxygen-free fermentation to oxidative phosphorylation, obtaining additional energy from oxygen. Aerobes were able to break down organic substances with greater efficiency than anaerobes. Some free-living airboats were captured by anaerobes, but were not "digested", and were preserved as power stations – *pro*-mitochondria, turned into MCs. To avoid terminological confusion, we will call the germinal MC, available in specialized animal cells, as *proto*-MC (PMC), but not pro-MC.

Endosymbiotic bacteria could develop into pro-MCs (Margulis et al. 1996). It was assumed that this union was unreliable, and therefore there were conflicts in the selection between the two genomes (Btackstone et al. 1995). Once the new symbiotic organism entered the world, its life and death were controlled by pro-mitochondria, which not only provided a number of necessary antioxidants,

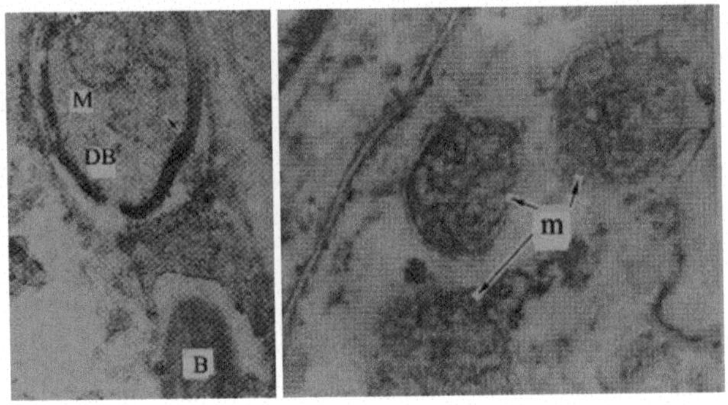

Fig. 2.6 Electron microscopy of the gill epithelium of mollusk (left) and fish (right). Notations: B is bacteria, introduced into the cell; DB is de-capsulated bacteria; M – mitochondria; m – mitochondria-like structure

but also serve as a source of superoxide, a byproduct of respiratory chain. Superoxide could lead to cell death and release of the endosymbiont, i.e. the symbiosis was initially unstable until the endosymbiont genes were transferred to the nuclear genome.

Many researchers are inclined to believe that the host cell belonged to archaebacteria (Margulis, Bermudes 1985; Vellai, Vida 1999). This is evidenced by the great similarity in the structure of the genome in archaebacteria and eukaryotes, the similarity of mechanisms of replication, reparation, transcription, translation, etc. (Cavalier-Smith 2002; Slesarev et al. 1998).

It is believed that all eukaryotes are monophyletic group, i.e. the fusion of archaebacteria with eubacteria into a single organism, which gave rise to eukaryotes, occurred in the history of the Earth only once (Gupta 1997). Genome analysis seems to have shown a common origin of all modern eukaryotes (Litoshenko 2002).

But there is another opinion that the various kingdoms of eukaryotes have various bacterial symbionts at different stages of evolution. This is evidenced by significant differences in the genomes of MCs in protozoa, fungi, plants and animals. But in all cases a part of genes from "endobacteria" was able to enter the nucleus.

The symbiotic hypothesis is supported by experimental results [21], in which the introduction of some modern bacteria into epithelial cells of mollusks and gills of fish was found (Fig. 2.6).

Although the hypothesis on symbiotic origin is almost universally accepted, separate voices against it continue to be heard (Cavalier-Smith 2002). Opponents of the symbiogenesis hypothesis have strong counter-arguments. First, mitochondrial DNA is only 16 kilo-bases, and bacterial DNA is millions. Second, MCs, unlike bacteria, usually not divide. Third, the penetration of bacteria into the cell is a process much more difficult than penetration of small DNA. Therefore, we cannot exclude that in the process of evolution in the primary cell was not infiltrated by bacteria, but – by DNA molecules. They could be implemented either (a) themselves, (b) like viruses - by injection from a phage. And then, in symbiosis with the cell, could develop into pro-MC, exploiting its nuclear genome.

In the book [18], it was emphasized that since MCs have own DNA, RNA and protein synthesis apparatus, they are fundamentally capable not only to independent life outside the cell, but even to transformation into microorganisms. In favor of this a set of data about strange transformation of isolated rat liver MCs into some microorganisms (not bacteria) was given. In a special environment, several dozen pieces of MCs (visible in the dark-field mode of microscope) in a few hours were transformed into several dozen pieces of microorganisms. Since the conditions were not completely sterile, it cannot be excluded that not MCs themselves turned, but some microbial spores sprouted on them, i.e. MCs served only as "food". But against this was the observation that the resulting microorganisms had the same small DNA of 16 kbs, according to electrophoresis. Verification experiments are required here.

2.4 Reproduction of Mitochondria in Cell

As already mentioned, the idea that MCs are autonomous self-replicating structures was proposed by Altman in 1890. Evidences for this hypothesis were obtained much later – in connection with the discovery of ability of MCs to store a number of functions in an isolated form (McLean et al.1958; Roodyn et al. 1961; Nass 1963; Rudin, Wilkie 1971) and in connection with the discovery of their own DNA (Nass 1963; Schatz et al. 1964; Kasamatzu et al. 1971) [22, 23].

The intracellular mitochondriogenesis has been debated for a long time. Three alternatives were considered: MCs are (a) formed by division, (b) from other membranes, (c) de novo.

The second and third hypothesis could be immediately discarded, because in this case there is nowhere to take 16-kilobase DNA. No intracellular organelle contains such DNA, and DNA cannot arise from anything.

As for the first hypothesis, it has no strict proofs. Nowhere in the literature, even in special monographs [22–27] there are no reliable facts. The textbook [28] provides a hypothetical scheme of dividing MC, invented by Muhlethaler in 1959, and some common words were said; but no facts.

Electron microscopic photos, where some MCs seem like divided, can be easy explained, on the contrary, their fusion or cross-linking by glutar-aldehyde (or other fixator used in the processing of electron microscopy sections).

Although each MC, having inside several copies of DNA, generally speaking, could be divided, but reliably microscopic observation of specialized animal cells shows, that it absent in the most cases. The division of MCs was observed (or allegedly observed) only in the case of cell cultures [29, 31–33, 35]. Moreover, the cells were exposed upon too strong influences, for example, powerful photoactivation of a dye was carried out [29], which could lead to both photochemical damage of the cell and to local heating that activates the ruptures. In [35], the fluorescent pictures in favor of the merger and division of MCs in stained HeLa cells were presented. However, these pictures can be interpreted by a different way. After all, the photos were obtained by two-photon microscopy, when photoprocesses and local thermal heating (through a dye) are very likely. It is interesting that the observed "worms" after the "division" did not move away far from each

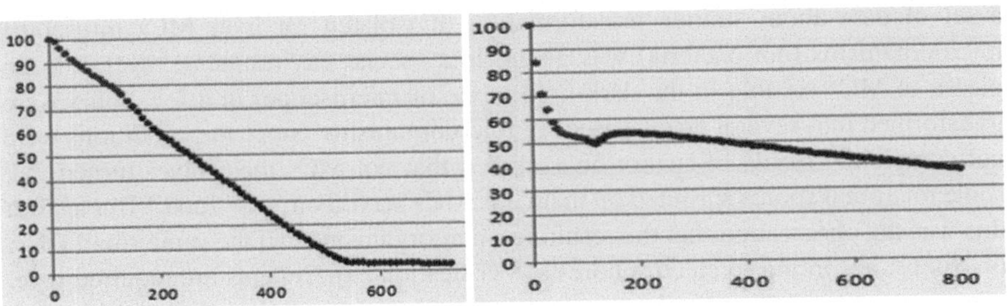

Fig. 2.7 Oxygen consumption on succinate during 10 min in suspension of yeast, grown in aerobic (left) and anaerobic (right) conditions (Vekshin N. L. Kurdyukov D. N., 2017)

other, i.e. small "constrictions" and "gaps" can be explained by the redistribution of dye, but not by division.

The monograph [30] presents photos of dark-field microscopy, where MCs in fibroblasts and specialized cells do not divide, but appear in cytoplasm like from nothing – de novo. It can be concluded that the size of the "embryos" is less than the length of the light half-wave, i.e. does not exceed 0.25 micron. But the question is, where does DNA come from into "embryos"? (answer will be given in final chapters of the book).

The hypothesis de novo, proposed at the beginning of the twentieth century, was based on the experiments of Beckwith (1914) and Harvey (1946). The formation of MCs de novo was also postulated by electron microscopic study of the ultrastructure of different cell types (Andre, Rouiller 1957, 1962) [22–25].

The question on the appearance of MCs de novo arose again in connection with the study of respiratory adaptation of yeast (Wallace, Linnane 1964). Analyzing anaerobically growing yeast cells with electron microscope, the researchers came to the conclusion that there are no MCs in them at all and, consequently, the transition of cells from anaerobic to aerobic conditions leads to the formation of MCs de novo (see the literature in [22, 23]).

Figure 2.7 shows the respiration of yeast, grown in aerobic and anaerobic conditions. In anaerobes, the respiratory chain does degrade: respiration on succinate (when cells are placed in the aerobic environment) is extremely low.

Unlike animal cells, the amount of MCs in yeast cells is very small, about ten pieces per one cell, so they are easy to follow. The optical microscopy photo of one dividing yeast cell is shown at Fig. 2.8. The nucleus of the cell is laced and divided into two parts (in this case unequal), but MCs are not divided; they are simply redistributed from "mother" to "daughter". Their number during cell division does not change. By the way, the clarity of the MC image here, without any staining, is much better than in [31–33], where fluorescence detection was performed when yeast were stained with "very specific" dyes. Staining can lead to a change in the shape of the supposedly MCs itself, their mobility and even division, although the diffusion of dyes was not taken into account for some reason, and the specificity of the dyes was not proved.

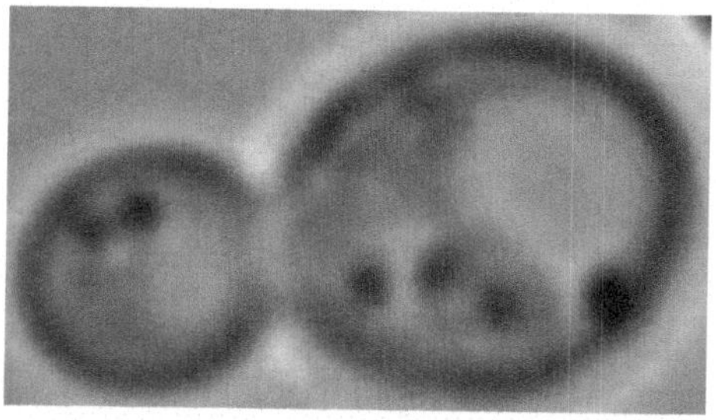

Fig. 2.8 Dividing of yeast cell, detected by optical quasi-dark-field microscopy. Dark points are MCs (Vekshin N. L. Kurdyukov D. N., 2017)

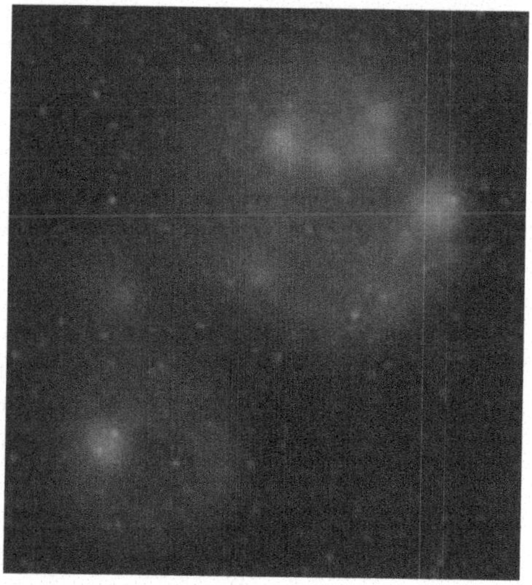

Fig. 2.9 Two yeast cells formed after dividing of one, detected by fluorescence microscopy (without staining). Bright spots are MCs (Vekshin N.L., Kurdyukov D. N., 2017)

Fluorescence of MCs in yeast is actually visible without any staining (Fig. 2.9). From Figs. 2.8 and 2.9 it follows that after the completion of cell division, the amount of MCs at first time remains the same.

And only after some time, the emergence of new MCs – either from small pro-MCs [22], or by the release of DNA from old post-MCs [34], or by budding (here requires verification) – begins.

MCs are dynamic structures and, in principle, are capable of merging and dividing. But the experimental data at the expense of division [29, 31–33, 35] is highly questionable.

The articles of Schatz et al. (1967–1972) about yeast under transition to anaerobic conditions showed, that MCs do not disappear completely, but strongly degrade, thereby becoming unobtrusive (see literature in [22]). In such particles there are no some enzymes, there is no complete respiratory chain; therefore, respiration is impossible [23]. When researchers return the yeast to aerobic conditions, these aberrant particles become the full-fledged MCs. Such aberrant particles in yeast were called pro-MCs. However, there was a terminological confusion, because many scientists previously called *pro-mitochondria* a hypothetical ancient “bacterial ancestor”, introduced into eukaryotic cells in the course of evolution and turned into MCs.

2.5 Do Mitochondria Swell Much?

One of the physiological reactions of MCs to intracellular stress may be their swelling [36, 37]. When isolated MCs are placed in a hypotonic medium (below 90 mOsm) or under the influence of fatty acids or other membrane-bonding compounds, the volume of each MC, as was suggested by many researchers, can supposedly increase many times. One of the main grounds for asserting such a high-amplitude swelling was the fact of a sharp decrease in the light scattering of the MC suspension [38]. A marked decrease in the light scattering of MCs can occur within a few minutes even in an isotonic environment, spontaneously. In the presence of EGTA (complexon on calcium), the effect is suppressed [39]. It was suggested that swelling occurs due to calcium-dependent phospholipase A2, located in MCs.

On the other hand, in papers [40–43] it was reported that saturated fatty acids, for example, palmitic acid (PA), can cause the formation of nonspecific pores in the lipid bilayer of MC membranes, regardless of the activation of phospholipase A2. However, similar pores arise in artificial phospholipid membranes in the absence of any phospholipase [43–45].

Both groups of authors [38, 39] and [40–43] interpreted the decrease in turbidity of the MC suspension as evidence that saturated fatty acids and calcium ions can cause strong swelling of each MC, accompanied by the ingress of a large amount of water inside. They identified the reduction in light scattering of MCs like high-amplitude swelling. This conclusion was done without direct morphological ultrastructural evidence. Moreover, the electron microscopy data presented in [43], in fact, do not reveal cardinal differences in the diameter of each MC before and after their processing using PA.

The statement about the high-amplitude swelling of MCs under the action of fatty acids, calcium, and other active substances stems from long-standing nephelometric, photometric, and electron microscopic studies of the middle of the last century (see monograph [38]). It was assumed that a strong decrease in light scattering of the MC suspension in response to the addition of PA or calcium is the result of high-amplitude swelling. But nowhere is there direct data from optical microscopy. Moreover, according to the laws of optics of turbid media [46, 47], with increasing particle diameter, light scattering should increase, but in the case of MC it decreases. That is why the researchers resorted to the assumption that each MC swells so much that a lot of water enters into matrix, as a result of which MC swells like a ball and becomes transparent.

Based on the laws of optics of turbid media [46, 47], a decrease in the light scattering of a suspension can occur for three reasons: (a) due to a decrease in the surface area of particles, (b) due to a decrease in the refractive index of particles, (c) due to a decrease in the number of particles.

It was necessary to verify the statement about the high-amplitude swelling of MCs using light scattering photometry and direct optical methods – phase contrast and quasi-dark field microscopy (fluorescence microscopy was not used here, because fluorescence probes bind differently to the outer and inner membranes and differently fluoresce outside and inside of native or damaged MCs [48]). It was important to establish the main causes of changes in light scattering in the suspension of MC under the influence of PA, calcium and hypotension.

MCs from rat liver were isolated according to the procedure described in [43], with modifications. Isolation was carried out at 4 °C. The liver was placed in 20 ml of ice-cold medium containing 210 mM mannitol, 70 mM sucrose, 0.5 mM EDTA (other option: without EDTA) and 10 mM HEPES (pH 7.4), after which the liver was pressed through a press. And then homogenization was carried out. The homogenate was centrifuged for 10 min at 1000 g. The supernatant was centrifuged for 15 min at 3000 g (fraction of heavy MCs) or at 6000 g (total fraction of MCs). The pellet was re-suspended in 8 ml of the same isolation medium. The MC suspension was aliquoted, stored on ice and used for experiments for several hours. The incubation medium for the optical microscopy experiments was used the same as isolation medium (without EDTA). The MC respiration in the other incubation medium (in the presence of 2 mM phosphate and 5 mM succinate) was detected by a polarographic Clarke electrode.

The light scattering of MCs was measured on a KFK-2 photometer by the optical density of the suspension in a 0.5-cm or 0.1-cm cuvette at 540 nm. Protein concentration was found at 286 nm by the UV method [49].

The overwhelming contribution (>95%) to the optical density D of the MC suspension, measured on a standard photometer (without a spherical detector) at 540 nm, is made by light scattering. The absorption of light by cytochromes can be neglected (see the MC spectrum in [47]). The optical light scattering density D detected by the photometer is described by Eq. 2.1 (similar to the known absorption law) [47]:

$$D = \sigma C\, L \tag{2.1}$$

Here, σ is the light scattering coefficient (optical scattering cross section), L is the optical path, C is the concentration of particles, which can be easily determined using a microscope.

Formula (2.1) is applicable for experiments on standard photometers and spectrophotometers (without light collection of scattered light by a sphere or hemisphere), which operate essentially in the mode of a turbodimeter or nephelometer – turbidity meters. The good applicability of formula (2.1) for the photometric determination of light scattering has been shown for various suspensions, in particular for MCs [47].

A spontaneous slow decrease in D of the MC suspension in the incubation medium (without phosphate and succinate) did not occur immediately, but after a short lag period, at the second minute (Fig. 2.10). This result differs from the kinetics from [39, 43], where there was no lag period. The presence of a lag period may indirectly indicate the gradual activation of the calcium-activated phospholipase pathway. Preliminary introduction of succinate (5 mM) into the incubation medium leads to a significant lengthening of the lag period (Fig. 2.10).

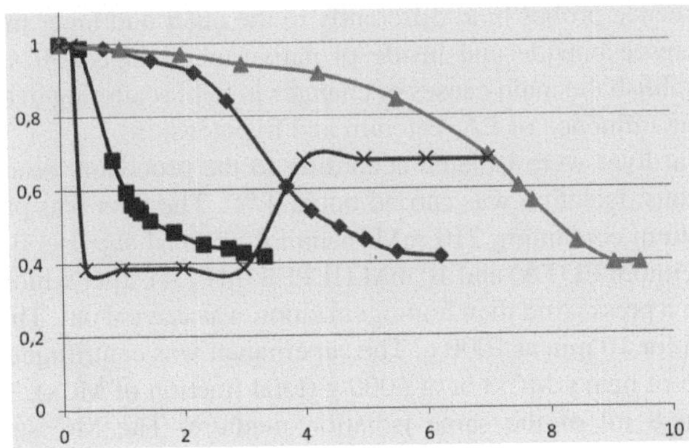

Fig. 2.10 Kinetics of changes in light scattering of a suspension of mitochondria in incubation medium without additives (□), in the presence of 5 mM succinate (▲), 30 μM palmitic acid (■), as well as mitochondria in water (□) supplemented by 60 μM calcium chloride at the third minute. The ordinate axis is the normalized optical density at 540 nm, the abscissa axis is the time in minutes. The incubation medium contained 210 mM mannitol, 70 mM sucrose and 10 mM HEPES (pH 7.4)

The addition of PA (or calcium chloride, see [38]) to the MC suspension on the background of succinate (or without it) was accompanied by a gradual fall of light scattering (Fig. 2.10), as in [38, 39, 43]. In the presence of EGTA or EDTA, no enlightenment from calcium or PA occurred (see [39]). EGTA and EDTA have a much higher affinity for calcium than PA. Therefore, endogenous calcium cannot be bound by PA in the presence of these chelators. Membrane permeabilization is caused not only by PA itself, but also by its complexes with endogenous calcium [43].

Earlier, some researchers suggested that the protective effect of succinate from the permeabilizing action of PA on MCs is supposedly due to the energization of the respiratory chain. However, it should be noted that complexons on calcium – EGTA and EDTA – have a similar effect. Since succinate has two negatively charged COO^- groups, it probably acts like EGTA or EDTA – it binds endogenous calcium. It prevents the damage of MC by calcium, released from matrix.

PA activates the succinate-dependent respiration of MCs (Fig. 2.11). It can be assumed that PA, creating defects in the lipid phase, increases the membrane permeability for succinate. Moreover, PA stimulates the succinate-dependent respiration of MC as much as calcium (Table 2.1, without succinate, PA and calcium do not affect respiration).

In all cases – isotonia, an in 80% hypotension, and after the addition of PA and even in the presence of succinate – the value of D eventually decreases almost the same – 2.5 times (Fig. 2.10). Hence, everywhere the fall of D is caused by the same optical reasons.

When calcium chloride is added to MCs (in isotonia), the D value decreases, but not so much. This is due to the fact that calcium cations, neutralizing the negative charge of MC, cause not only permeabilization of membranes, but also partial aggregation of MCs, which is accompanied by the opposite effect – increased

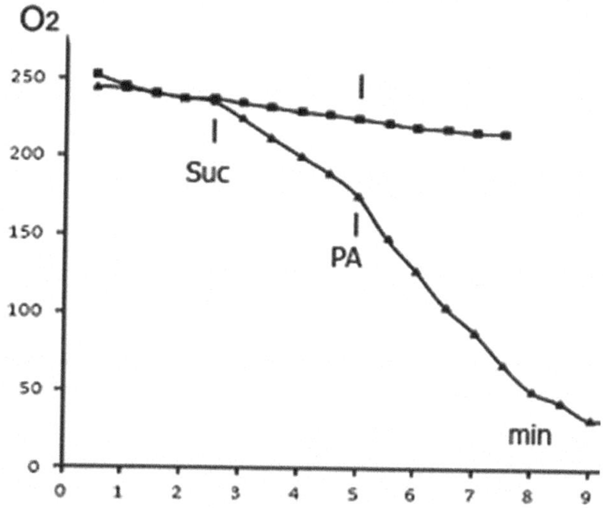

Fig. 2.11 Oxygen consumption in suspension of mitochondria (1 mg/ml) with the addition of 5 mM succinate and then 60 μM palmitic acid (▲) and with the addition of only palmitic acid (■). The incubation medium contained 2 mM phosphate, 210 mM mannitol, 70 mM sucrose and 10 mM HEPES (pH 7.4)

Table 2.1 The rate of oxygen consumption by suspension of mitochondria (1 mg/ml)

	Mitochondria					
	Without additions	60 μM $CaCl_2$	30 μM PA	5 mM succ.	5 mM suc. + 60 μM $CaCl_2$	5 mM suc. + 30 μM PA
Respiration rate O_2 (μM/min)	5.4	5.5	5.5	24	46	47

scattering. Aggregation on the calcium addition is especially noticeable in complete hypotension (water), when other cations and anions do not interfere with calcium binding on the membrane surface (after adding calcium in the third minute, D increases, Fig. 2.10). Large aggregates with a size of 5–10 microns, apparently, were mistaken interpreted by some authors as huge swollen MCs.

When MCs are placed in distilled water, the suspension enlightens instantly – just of a second (Fig. 2.10), which makes the phospholipase mechanism unlikely, because it takes time to hydrolyze lipids with a phospholipase. Under hypotonic conditions, high osmotic pressure from the inside leads to membrane damage. The outer membrane breaks and partially flies into the water [36, 37]. This should lead to a decrease in light scattering, because under these conditions, the reflection and refraction of the light beam will occur on only one membrane, unlike whole MC, where light is scattered directly on two membranes – external and internal.

The probability of scattering of photons (p) by single MC at a fixed wavelength can be represented as the ratio of two areas [47]:

$$p = \sigma / \pi R^2 \tag{2.2}$$

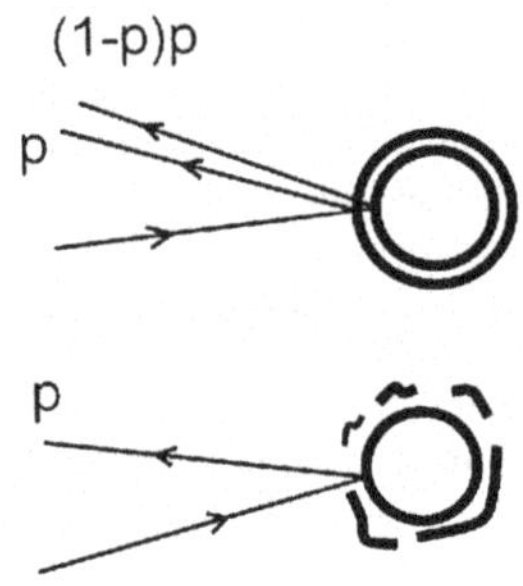

Fig. 2.12 Scheme of light scattering on membranes of native and damaged mitochondria

Here πR^2 is the transverse area of one MC. According to formula (2.2) for a suspension of hepatic MCs, it was previously found that $p = 0.32$ [47] (at 540 nm), i.e. 32% of the photons incident on a single MC are scattered to the sides.

The probability of light scattering on a single particle in diluted suspensions does not depend on the concentration of particles. Besides, it does not depend on the detection method. In a spectrophotometer with a spherical detector, the light scattering of the suspension is collected, but the scattering of particles themselves cannot be compensated. It must be understood that even with strong dilution of the particles, the absorption spectra (for example, the spectra of tryptophan residues, flavins and cytochromes) will be hypochromized and smoothed, in contrast to the spectra of ideal solutions.

The contribution to the total MC scattering is provided by the water/membrane, membrane/matrix, and water/matrix phase boundaries (Fig. 2.12). The refractive index of water is 1.33, the membrane is about 1.6, and the matrix is about 1.8.

Let the probability of photon scattering on the outer membrane be equal to P_{out}. Then for the inside membrane it is smaller, namely $(1 - P_{out})\ P_{in}$, where P_{in} – probability without taking into account the outer membrane. Since the inter-membrane distance does not exceed 20 nm, and the thickness of the outer membrane is 10 nm [37], the "outer" refractive index of the inner membrane will be something average between 1.33 and 1.6, so P_{in} should be less than P_{out}. But since the inner membrane is many-folded, this should vice versa increase the value of P_{in}. Since these two factors are multidirectional, for simplicity we can take, as a first approximation, $P_{out} = P_{in} = P$. Then the total probability of photon scattering on two membranes will be $2P - P^2$. As already mentioned, 32% of photons scattered from MC to the sides, i.e. $2P - P^2 = 0.32$. From the solution of this equation it follows that $P = 0.18$. The ratio of the total probability of scattering on two membranes to the probability of scattering on one membrane is $(2P - P^2)/P$. It gives 1.8. Thus, the value of D after removal of the outer membrane or (and) leakage of the matrix should decrease about 1.8 times (within the framework of the assumptions made).

A significant contribution to the decrease in D can be made by the yield of substances from the matrix, since its refractive index is higher than water. According to electron microscopy, in the matrix of each rat liver MC, in addition to cristae of the internal membrane, there are dark spots – places of increased electron density corresponding to the accumulation of calcium phosphate granules [37], which can make a significant contribution to light scattering.

The Tyndale light scattering of suspensions, when the size of each particle is significantly larger than the wavelength λ, depends on the refractive index, particle size, and wavelength [46, 47], i.e. $\sigma \sim n R/\lambda$. Rayleigh molecular light scattering sharply increases with the aggregation of molecules or with an increase in n, namely, $\sigma \sim n^4 R^4/\lambda^4$. In the intermediate case, when the particle size is commensurate with the wavelength (MC has diameter ~ 1 μm, λ = 0.54 μm), light scattering can be approximately described by a quadratic dependence

$$\sigma \sim n^2 R^2 / \lambda^2 \tag{2.3}$$

When MCs are incubated in isotonia or hypotension, flavins [50], calcium [39], as well as some proteins [37, 38] spontaneously rapidly leave their membranes and matrix, as a result of which n can decrease from ~ 1.6 (granules of calcium phosphate, proteins and DNA) to 1.33 (water), i.e. 1.2 times. It follows from formula (2.3) that in this case, σ will decrease by 1.44 times. Therefore, according to formula (2.1), D will decrease by the same amount.

If both reasons are taken into account (one membrane instead of two and emptying the matrix), the drop in D has to be 2.6 times. This value is in good agreement with experiment – 2.5 times (Fig. 2.10).

In articles [51, 52], where the presence of high-amplitude swelling was stated, the probable dependences of the optical density of the MC suspension on the matrix volume were considered using several indirect methods, but optical microscopy was not used.

Electron microscopic data on extremely strong MC swelling in hypotension were presented in the article [53], but they raise doubt: the diameter of each MC when it enters the water increased supposedly 5–6 times; this means that the volume increased by at least a hundred times, which is difficult to imagine and which was not subsequently observed in any electron microscopy study (all changes in the diameter of MCs do not exceed 20–40%).

It should be emphasized that electron microscopy deals with fixed, non-native samples, and the use of glutar-aldehyde or formaldehyde in the manufacture of slices could stitch aggregated MCs together (see this effect of glutar-aldehyde on proteins in [48]), creating the illusion of giant MCs.

In [54–56], it was argued that changes in the light scattering of MCs are caused by changes in the internal volume. But there is no data from native optical microscopy. The optical microscopy method was also not used in [57–59].

In this regard, experiments using optical microscopy were carried out [60]. Particle size was determined by quasi-dark-field and phase-contrast optical microscopy by LUMAM-I2 and Leica DM-6000. The average particle size was obtained from 5–10 microphotographs using the ImageJ program.

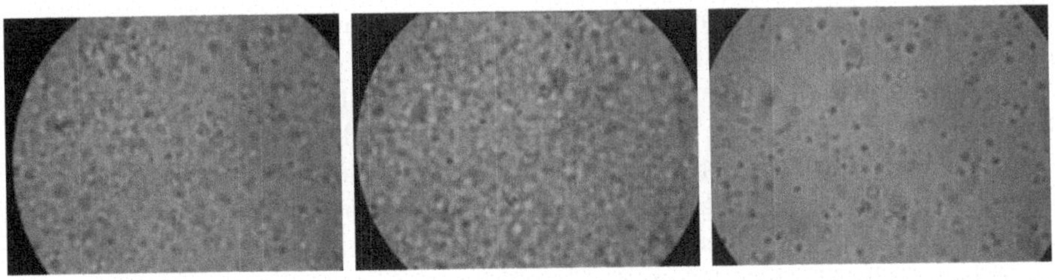

Fig. 2.13 Mitochondria (heavy fraction) immediately after their addition to the medium (left), after 3 min of incubation in medium with palmitic acid (in the center) and in a hypotonic medium obtained by fivefold dilution with water (right). Incubation medium was at the presence of 5 mM succinate. LUMAM-I2, quasi-dark-field regime, water immersion, magnification ×2250

Table 2.2 Quasi-dark-field microscopy of heavy mitochondria

Single MC	Initially	With PA	In water
Average diameter (μm)	1.03	1.14	0.94
± (μm)	0.23	0.22	0.15

Note: The isolation medium did not contain EDTA. Diameter of a single MC is the arithmetic mean of 200–300 particles

The contours of each MC under microscopy in a quasi-dark-field mode at maximum magnification (×2000) are close to the contours in the phase-contrast mode.

Figure 2.13 shows the microscopy of a suspension of a heavy fraction of hepatic MCs in a quasi-dark-field mode initially and after incubation with PA, as well as MCs in 80% hypotension. No high-amplitude changes in the size of MCs are observed. No wonder. The outer membrane is damaged, but the volume of the MC remains the same, because the inner membrane is not at all like a rubber ball that can swell.

Judging by quasi-dark-field microscopy, with 80% hypotension, the average diameter of a single MC even slightly decreases (Fig. 2.13c and Table 2.2). Apparently, this is due to the removal of the outer membrane. If we talk about MC swelling under the action of PA (Fig. 2.13b and Table 2.2), then it's not at all about high-amplitude, but about miserable, insignificant.

Figure 2.14 shows microphotographs of the total MC fraction in phase-contrast mode (a) initially, (b) after incubation for 37 min at room temperature, (c) after incubation with PA for 20 min, (d) in 100% hypotension (distilled water) after 2 min. No high-amplitude changes of MCs are observed either in the presence of PA, or even in complete hypotension. Only after 12 min in water does the diameter increase by 42% (Table 2.3) and enlightenment of each MC occurs. Moreover, in the case of incubation with PA and without, the volume of each MC did not noticeably change. A small (by 16–22%) increase in diameter cannot be the cause of a 2–3-fold drop in turbidity, as measured by a decrease in optical density.

So, there is reason to believe that hypotension or induction of nonspecific pores in MC by calcium and PA does not lead to strong expansion of the inner membrane and

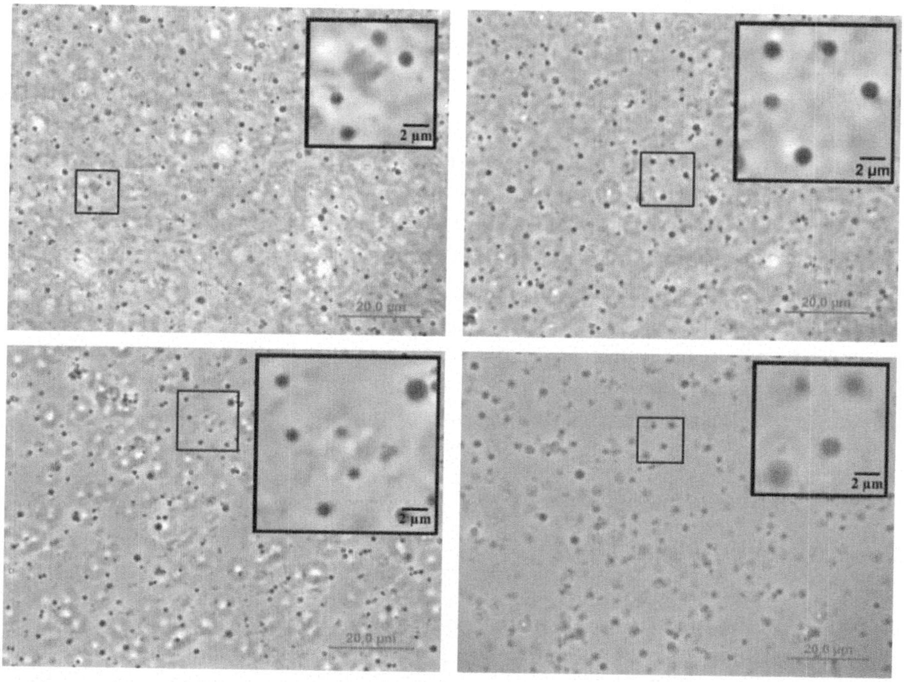

Fig. 2.14 Mitochondria (total fraction) immediately after their addition to incubation medium (top left), after 37 min (top right), after 20 min of incubation with palmitic acid (bottom left) and after 2 min in water (bottom right). Leica-DM-6000 microscope, phase contrast, oil immersion. An enlarged scale is shown at the insets

Table 2.3 Phase-contrast microscopy of the total mitochondrial fraction

MCs	Isotonia, initially	Isotonia after 37 min	Isotonia + PA after 20 min	Water after 2 min	Water after 12 min
Average diameter (microns)	0.72	0.84	0.88	0.88	1.02
Diameter in %	100	116	122	122	142

Note: The isolation medium contained 0.5 mM EDTA. Diameter of one MC is arithmetic mean of 400–2000 particles. Leica DM 6000 microscope, phase contrast, oil immersion

to high amplitude swelling of MC, but cause damage of outer membrane (up to destruction) and increase the permeability of the inner membrane, as a result of which the matrix is empty.

In general, it has long been noted [59] that incubation of isolated MCs alone usually leads to the release of pyridine nucleotides, proteins, etc. from them.

In conclusion, the biomedical aspect should be noted. High PA concentrations damage the MC. This means that the constant use in human nutrition of large amounts of palm oil and other oils and products containing many free fatty acids can cause great harm to the human body (we are talking primarily about free PA or its complexes with calcium, but not PA which is part of triglycerides).

Conclusion In contrast to the claims of many authors about the high-amplitude swelling of MCs under the influence of palmitic acid, calcium ions or hypotension, it was shown that MCs are not subject to strong swelling. The decrease in light scattering of the MC suspension in hypotension or with the addition of palmitic acid or calcium (in isotonia) occurs due to damage of the outer membrane and an increase in the permeability of the inner membrane, leading to the emptying of the matrix. Palmitic acid and calcium cause permeabilization and destruction of the outer membrane gradually, over several minutes. At low ionic strength, the addition of calcium, which neutralizes negative charges on the surface of the membranes, initiates the aggregation of MC, which, on the contrary, enhances light scattering and can create the illusion of their swelling. The results were published in [60].

2.6 Mitochondrial DNA

In 1909, Corrence and Baur noticed that some phenotypic traits are inherited only via the maternal line, and, this heredity was not nuclear, but cytoplasmic (T. Morgan 1924). Signs encoded by MCs were transmitted in animals for at least of 25 generations (Michaelis 1924) [61, 62].

Comparative analysis of mitochondrial DNA (by type, location and number of individual mutations), taken from 150 representatives of ethnic groups of different races, showed that all DNA originated from one ancestor (A. Wilson 1987). All of humanity came from one woman – mitochondrial Eve, who lived in North-East Africa more than 100 thousand years ago. Mitochondrial DNA is inherited through the maternal line [63].

DNA in MC is a closed ring double-helical macromolecule (Fig. 2.15) with a mass of about 16 kDa. In humans, it consists of 16,569 nucleotide pairs. Unlike nuclear DNA, the mit-DNA almost "naked", not associated with histones [64], but usually very crowded.

Human mitochondrial DNA is a hundred times smaller than nuclear DNA. It encodes 2 rRNA, 22 tRNA and 13 subunits of respiratory chain enzymes (Fig. 2.16). The main part of the subunits and proteins are encoded by nuclear DNA.

Mitochondrial DNA replication in the inter-phase can be synchronized with nuclear DNA replication.

Each MC in an animal cell has its own protein-synthesizing system with small ribosomes, having a sedimentation coefficient of 55S.

Single MC has several copies of DNA; in mammals – up to ten pieces in each. This makes possible to choose "good" copies, but "bad" (oxidized, methylated) – to dispose of, that is very important for those MCs, which are not able to restore DNA from damages (high mutation rate). Mutations in mitochondrial DNA are the cause of a number of hereditary diseases.

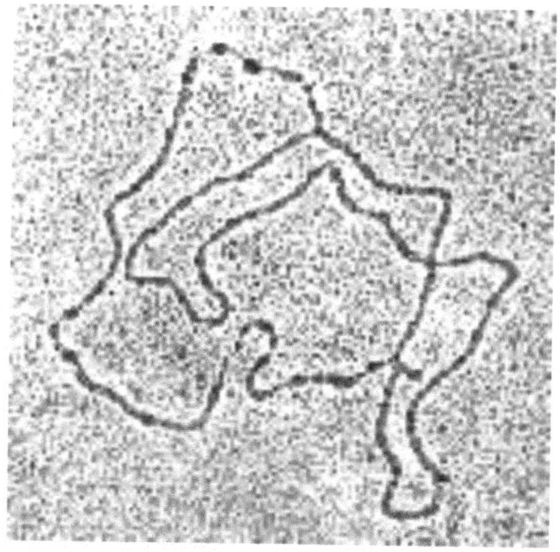

Fig. 2.15 Electron microscopy of un-twisted mitochondrial DNA. http://www.ggause.com/images/mtdna04.jpg

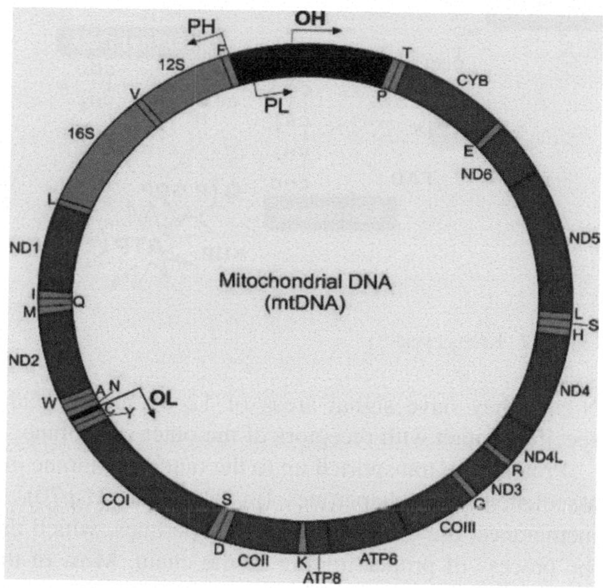

Fig. 2.16 Map of mitochondrial DNA genes. Genes encoding subunits of ATPase, NADH dehydrogenase, cytochrome B and cytochrome oxidase are indicated

2.7 Membrane's and Matrix's Proteins

According to the "MitoCarta", the composition of MC consists of not less than 1100 various proteins [65].

In the inner membrane, 20% of proteins are red-ox and enzymes, and 80% are structural proteins [66]. The structural proteins form the inner membrane, with small inclusions of respiratory ensembles and lipids.

The majority of MC proteins, not less than 85%, in particular – cytochromes, are encoded in the nucleus and synthesized on cytoplasmic 80S ribosomes [67]. At its

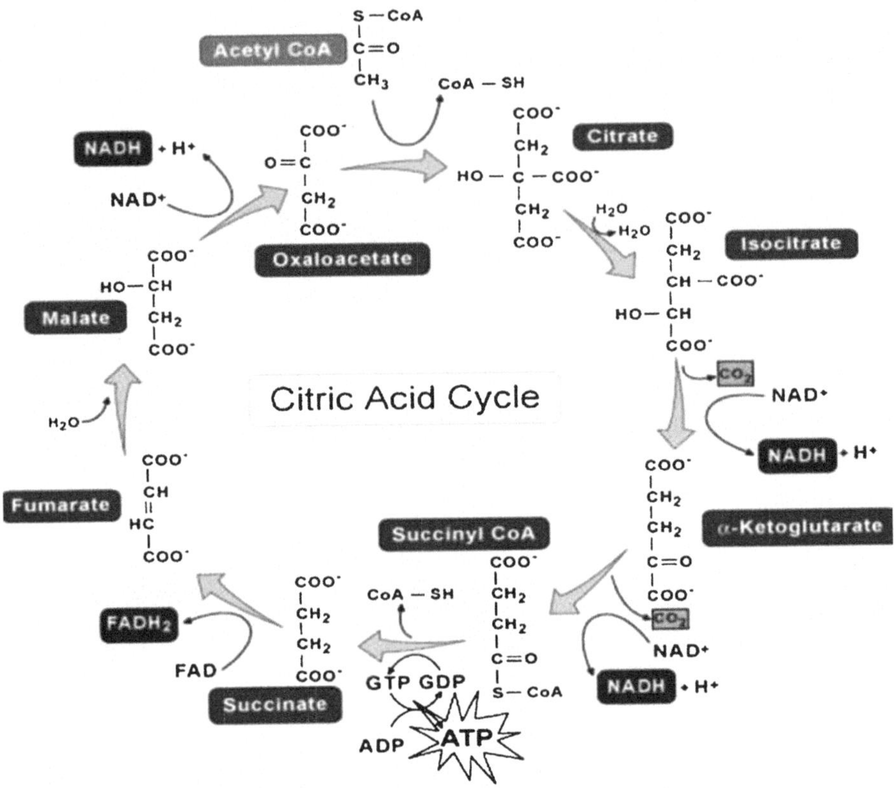

Fig. 2.17 Krebs cycle

N-end, they have signal areas of 12–80 amino acids, forming curls, providing specific contact with receptors of the outer membrane.

Proteins are transported up to the outer membrane in a partially unfolded state in association with chaperones (in particular, Hsp70). After transfer through two membranes, they bind to matrix chaperones, which drag them inside and control the process of proper folding of the chain. Most of the chaperones have ATPase activity. It means that they can use ATP as a source for working.

Matrix contains enzymes of Krebs cycle [66, 67] – tricarboxylic acid cycle (Fig. 2.17). Acetyl-coenzyme comes from outside to matrix and turns oxalo-acetate to citrate and then iso-citrate, which is decarboxylated to keto-glutarate with the reduction of NAD to NADH. Keto-glutarate is decarboxylated and dehydrated, restoring the second NAD molecule. Received succinyl-coenzyme releases a coenzyme A that is accompanied by substrate phosphorylation. Succinate is dehydrogenated with FAD reduction, and the resulting fumarate turns into malate, which is dehydrogenated with the reduction of the third NAD molecule. The resulting oxalo-acetate reacts with acetyl-coenzyme A.

NADH and succinate then are oxidized in the respiratory chain, giving a lot of energy in the reduction of oxygen to water.

Thus, during the Krebs cycle functioning, dehydrogenation of di – and tricarboxylic acids occurs within the reduction of FAD and NAD to $FADH_2$ and NADH, as well as non-membrane synthesis of ATP. Moreover, decarboxylation of ketoglutarate can be carried out non-enzymatically, without any participation of proteins, for example, with the participation of hydrogen peroxide [68].

2.8 Bioenergetics

The energy formation in MC has four stages; the first two occur in matrix, and the last two – on cristae: (1) conversion of pyruvate and fatty acids (from cytoplasm) to acetyl-coenzyme-A; (2) oxidation of acetyl-CoA in the Krebs cycle with restoration of the pyridine ring of NAD (Fig. 2.18), as well as with release of two CO_2 molecules; (3) electron transfer from the formed NADH to oxygen through respiratory chain, with appearance of H_2O at cytochrome oxidase; (4) synthesis of ATP by the membrane ATP-synthetase complex.

MCs are commonly called "intracellular power stations", because one of their main functions is the extraction of energy from organic compounds (substrates) and oxygen, which oxidizes them [69].

It should be emphasized that the energy is extracted not only from substrates, but also directly from oxygen, which, after attaching protons and electrons, gives water molecule that accompanied with release of noticeable energy:

$$½\, O_2 + 2H^+ + 2\,e \rightarrow H_2O \qquad \Delta G = \ -18.7\ kcal/mol$$

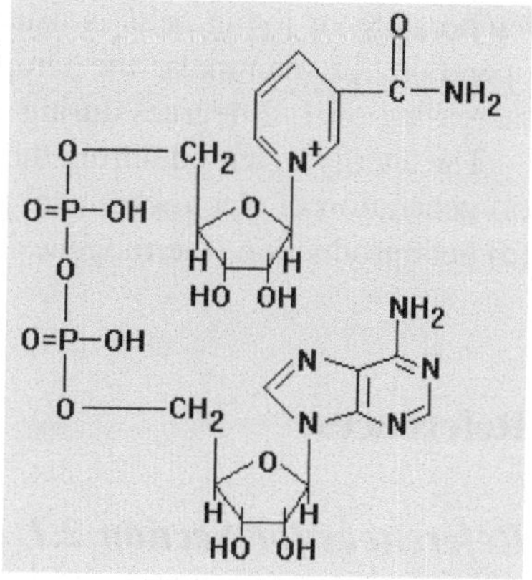

Fig. 2.18 Structural formula of nicotinamide adenine dinucleotide (NAD)

Such significant energy (ΔG is standard Gibbs free energy at concentrations = 1 M and pH = 7) in the aerobic process is released mainly as a result that molecular oxygen, which is a triplet with two unpaired electrons, has much higher energy level than water molecule with paired singlet electrons (oxygen on Earth arose as a result of photosynthesis, in which the energy of photons is stored not only in chemical bonds of organic molecules, but also in the resulting oxygen).

Under anaerobic processes, much less energy is being released, than under aerobic processes. For example, dehydrogenation of succinate to fumarate:

$$Succinate \rightarrow Fumarate \qquad \Delta G = -0.73\ kcal/mol$$

Due to that the Krebs cycle provides repeated dehydrogenation (and twice – decarboxylation) of di-/tricarboxylic acids, the balance is dialed up to −14.8 kcal/mol – a value commensurate with aerobic process, but still smaller.

It is due to oxygen the process of glucose utilization to carbon dioxide and water gives 14.5 times more energy than the splitting of glucose to lactate:

$$Glucose + 6\ O_2 \rightarrow 6\ \mathrm{CO}_2 + 6\ H_2O \qquad \Delta G = \quad -686\ kcal/mol$$

$$Glucose \rightarrow 2\ Lactate \qquad \Delta G = \quad -47\ kcal/mol$$

With the inclusion of intracellular respiration, a decrease in glucose consumption takes place and the cessation of lactate formation is detected (the effect of Pasteur, for example, in yeast).

NADH oxidation in respiratory chain gives in a balance ΔG = −52.7 kcal/mol. There are a lot of NADH-dependent enzymes in cells. Thus, NADH in cells is a powerful source of energy.

Intracellular respiration is an enzymatic process: the oxidation of substrates in MC is catalyzed by special enzymes combined in respiratory chain. The respiratory chain "burns" the substrates with oxygen by the same way as gasoline burns in a car engine. Unlike the engine where the temperature reaches two thousand degrees, the temperature of living cells is usually 30–40 degrees and few increases during the operation. In mammals, the temperature inside the organs is about 38 degrees, increasing on 1–2 degrees during active work.

The energy, released during the functioning of respiratory chain, can be spent for (a) generation of electrochemical potential, (b) synthesis of ATP, (C) ion transport, (d) heat production (thermogenesis) [70].

References

References for Section 2.1

1. Lehninger A (1974) Biochemistry. Mir, Moscow

2. Belyakovich AG (1990) Study of mitochondria and bacteria using the salt tetrazolium pNTV. ONTI, Pushchino
3. Cox M, Nelson D (2017) Principles of Lehninger's biochemistry, vol 2
4. Vekshin NL (1988) Biophysics 33(2):360–362

References for Section 2.2

5. Brandt JT et al (1974) BBRC 59(3):1097–1113
6. Bakeeva LE, Skulachev VP, Chentsov YS (1977) Vestnik Moskow Univ 3:23–38
7. Johnson LV, Walsh ML et al (1980) PNAS 77(2):990
8. Kulik AV, Nekrasova OE, Minin AA (2006) Biol Membr 23(1):42
9. Belyakovich AG (1990) Study of mitochondria and bacteria using tetrazolium salt pNTV. ONTI, Pushchino
10. Vekshin NL, Frolova MS (2014) Biophysics 59(5):1034–1039
11. Chentsov YS (2004) Introduction to cellular biology. Akademkniga, Moscow
12. Vekshin NL (2008) Fluorescence spectroscopy of biopolymers. Photon, Pushchino
13. Vowinckel J et al (2015) Mitochondrion 24:77–86
14. Chernyak BV (2010) Diss Cand Doctor Biol Sci. Moscow State University
15. Frolova MS, Chachina NA, Kirtok AN, Vekshin NL (2013) In: Neuroinformatics, part 1. Moscow, pp 219–229
16. Zinina AN, Vekshin NL (2008) Biol Memb 25(6):480–487
17. Vekshin NL (2010) Biol Membr 27(5):424–429

References for Section 2.3

18. Belyakovich AG (1990) Study of mitochondria and bacteria using the tetrazolium salt pNTV. ONTI NCBI, Pushchino
19. Alberts B (2013) Molecular biology of cell. Moscow
20. Cox M, Nelson D (2017) Principles of Lehninger's biochemistry
21. Kuznetsov AP, Lebkova NP (2002) Izvestiya of Russ Acad Sci Ser Biol. 4:501–507

References for Section 2.4

22. Rudin D, Wilkie D (1970) Biogenesis of mitochondria. Mir, Moscow
23. Kotelnikov AV, Ostapchuk RA (1973) Biochemistry of yeast mitochondria. Nauka, Moscow
24. Ozernyuk ND (1978) Growth and reproduction of mitochondria. Nauka, Moscow
25. Luzikov VN (1980) Regulation of mitochondrial formation. Nauka, Moscow
26. Alberts B (2013) Molecular biology of cell. Mir, Moscow
27. Cox M, Nelson D (2017) Principles of Lehninger's biochemistry
28. Chentsov Yu S (2004) Introduction to cellular biology. Akademkniga, Moscow, pp 324–355
29. Twig G et al (2008) EMBO J 27:433–446
30. Belyakovich AG (1990) Study of mitochondria and bacteria using tetrazolium solt pNTV salt. Pushchino
31. Vowinckel J et al (2015) Mitochondrion 24:77–86

32. Li T et al (2015) Sci Rep 5:11023
33. Friedman JR et al (2011) Science 334(6054):358
34. Begunova EA, Vekshin NL (2015) Biophysics 60(6):1109–1117
35. Wang S et al (2012) PLoS One 7(5):e19879

References for Section 2.5

36. Alberts B (2013) Molecular cell biology, vol 2. Moscow
37. Lehninger A (1974) Biochemistry. Mir, Moscow
38. Lehninger A (1966) Mitochondria. Mir, Moscow
39. Olenev VI, Suslova TB, Vladimirov YA (1976) Stud Biophys 2:147
40. Bernardi P, Petronelly V (1996) J Bioenerg Biomembr 28:131
41. Crompton M (1999) Biochem J 341:233
42. Bernardy P (1999) Phisiol Rev 79:1127
43. Belosludtsev KN (2005) Diss Cand Biol Sci. Pushchino
44. Antonov VF et al (1976) Free radical oxidation in norm and pathology. Nauka, Moscow
45. Mironova G et al (2001) J Bioenerg Biomembr 33:319
46. Lopatin VN (2004) Light scattering methods in the analysis of dispersed biological media. Fizmatlit, Moscow
47. Vekshin NL, Frolova MS, Kovalev VI, Begunova EA (2015) Biophysics 60:129
48. Vekshin NL (2006) Fluorescence spectroscopy of biopolymers. Pushchino
49. Vekshin NL (2002) Photonics of biopolymers. Springer, Berlin
50. Frolova MS, Vekshin NL (2014) J Fluoresc 24:1061
51. Tedeschi H, Harris D (1955) Arch Biochem Biophys 58:52
52. Tedeschi H, Harris D (1958) Biochim BBA 28(2):392
53. Stoner C, Sirak H (1969) J Cell Biol 43(3):521
54. Beavis A et al (1985) J Biol Chem 260(25):13424
55. Garlid K, Beavis A (1985) J Biol Chem 260(25):13434
56. Gotterer G, Thompson T, Lehninger A (1961) J Biophys Biochem Cytol 10:15
57. Massari S (1996) J Biol Chem 271(50):31942
58. Reiss M, Costa A (2008) Anastes Analges 106(4):1049
59. Luzikov VN (1980) Regulation of formation of mitochondria. Science, Moscow
60. Kurdyukov DN, Vekshin NL (2016) Biophysics 61(4):736–743

References for Section 2.6

61. Lehninger A (1966) Mitochondria. Mir, Moscow
62. Jinx J (1966) Non-chromosomal inheritance. Moscow
63. Igamberdiev AU (2000) Soros Educ J 6(1)
64. Cox M, Nelson D (2017) Principles of Lehninger's biochemistry, vol 1–3

References for Section 2.7

65. Kosterina EA (2013) Bull Sib Med 12(3):5–17
66. Lehninger A (1974) Biochemistry. Mir, Moscow
67. Cox M, Nelson D (2017) Principles of Lehninger's biochemistry, vol 1–3
68. Fedotcheva NI, Sokolov AP, Kondrashova MN (2006) Free Radic Biol Med 41:56–64

References for Section 2.8

69. Lehninger A (1974) Biochemistry. Mir, Moscow
70. Skulachev VP (1985) Stories about bioenergetics. Moscow

Chapter 3
Structure and Function of Respiratory Chain

3.1 Respiratory Chain

Each respiratory chain is a “breathing ensemble”. One hepatic MC contains about 5000 ensembles.

Figure 3.1 shows the conventional scheme of proton and electron transport during the oxidation of NADH in respiratory chain of the inner membrane.

In this common-accepted scheme, there are a number of inaccuracies: (a) the length of the lipid phase is greatly exaggerated; in fact, the membrane mainly consists of protein, with lipid inclusions, (b) there are no contacts between the complexes; but the rapid transfer of electrons through lipid phase is impossible, since lipids are insulators, (c) oxygen is taken by cytochrome oxidase allegedly from aqueous phase; but it must be taken from the lipid phase, because its solubility in lipids is 4 times higher, (d) when NADH is oxidized, one proton seems returned to matrix, i.e. a part of energy is wasted that is strange.

The transfer of electrons along respiratory chain occurs due to the gradient of red-ox potentials of its components (Fig. 3.2). The value of oxidation-reduction potentials and the released energy are usually calculated for two-electron transport, rather than one-electron, although cytochromes b and c, unlike other red-ox carriers, can actually take only one electron. In the case of one-electron transport, too little energy is released in the first two interface points, i.e., there, according to conventional views, ATP could not be synthesized (but it is very controversial). With two-electron transfer in two points of energy seems to be enough, but in the third point it is too much, since 70% is “lost” as a heat.

In fact, two-electron transport mechanism is indeed required, but for quite another reason: for the reunification at least of two electrons and 1–2 (or 3–4) protons from

N. Vekshin, *Biophysics of Mitochondria*,
https://doi.org/10.1007/978-3-030-33853-4_3

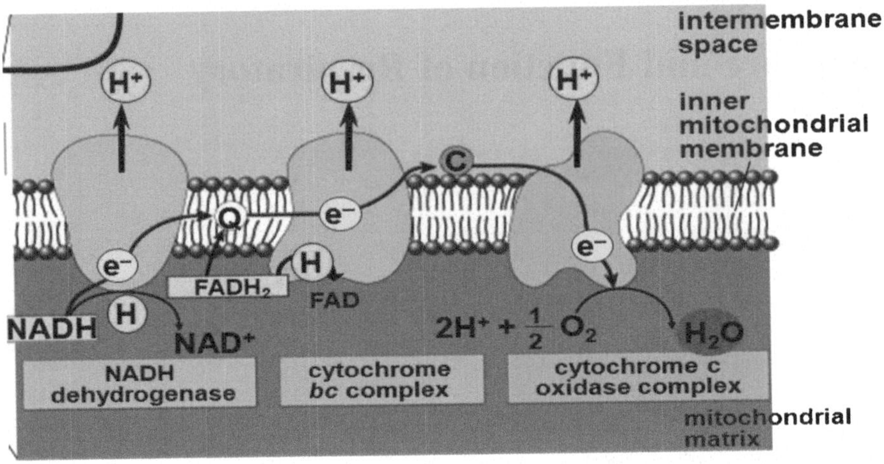

Fig. 3.1 Conventional scheme of proton and electron transport during oxidation of NADH in respiratory chain

the substrate to cytochrome oxidase (NADH gives 1H^+, succinate gives 2H^+) with molecular oxygen (a single-electron mechanism is not possible here):

$$2H + O_2 \rightarrow H_2O_2 \qquad \text{or} \qquad 4H + O_2 \rightarrow 2H_2O$$

The sequence of red-ox centers was invented about 60 years ago on the basis of standard red-ox potentials [1]. But it has no direct relation to the reality, since real concentrations are not equal to 1 M and pH in matrix differs from 7. As for the experimental evidence, the sequence was based on not too correct interpretation by Chance [2] and other researchers of spectroscopic data for turbid suspensions of MCs and cells (see critical analysis in [3]).

Although the scheme in Fig. 3.2 is very contradictional, but qualitatively it correctly conveys the essence: the transfer of electrons between red-ox centers is accompanied by the release of energy sufficient for the appearance of at least three ATP molecules. Points of coupling between electron transfer and ATP synthesis (Fig. 3.3) are hypothetical; they are generally accepted but not strictly proven.

Electron transfer along the chain consists of several stages, each of which involves transmembrane complexes: NADH dehydrogenase complex I, b-c-cytochrome complex III and cytochrome oxidase complex IV (Figs. 3.1, 3.2 and 3.3). Electron transfer from succinate is carried out by succinate dehydrogenase complex II.

At the time during the NADH oxidation, the transmembrane transfer of H^+ occurs, resulting in electrochemical proton gradient. And two electrons from NADH are transferred along the chain to cytochrome oxidase and then to oxygen. In many schemes, the transmembrane transfer at the level of complex I was assumed not as single-proton transport, but the transport of two or even four protons. But the

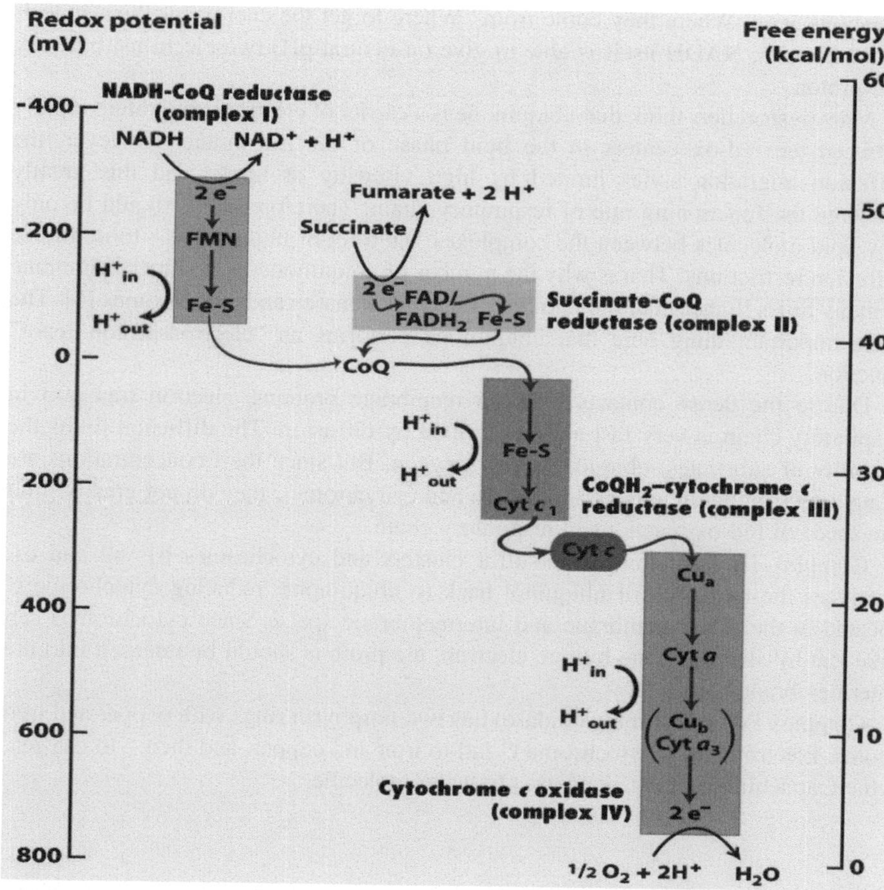

Fig. 3.2 Standard red-ox potentials (eV) and energy (kcal/mol) at pH = 7 for 2-electron transport along the respiratory chain

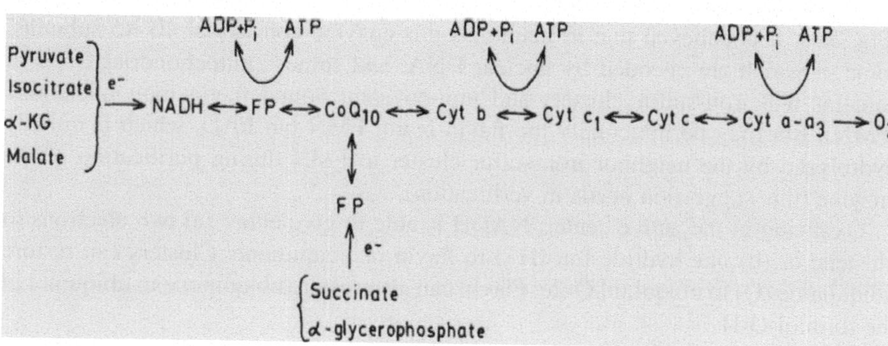

Fig. 3.3 Complexes I–IV of respiratory chain and the sites of coupling with ATPase

questions arise. Where they come from? Where to get the energy for their appearance? Besides, NADH itself is able to give (at neutral pH) two electrons, but only one proton.

Many researchers think that ubiquinone is a carrier of electrons, migrating rapidly between the red-ox centers in the lipid phase of the membrane. However, the diffusion migration styles limited by high viscosity of lipids, and this greatly hindered the functioning rate of respiratory chain. Therefore, there should be only few lipid molecules between the complexes, but a lot of ubiquinone – to minimize diffusion restrictions. That is why the number of ubiquinone in the inner membrane is many times higher than the number of dehydrogenases and cytochromes [4]. The most important thing here that ubiquinone performs an "electron-proton-depot" function.

Due to the dense contacts between membrane proteins, electron transport in respiratory chain is very fast and not limited by diffusion. The diffusion limits the mobility of substrates, ubiquinone and oxygen. But since their concentrations are many times higher than dehydrogenases and cytochromes, they do not greatly limit the speed of red-ox reactions in respiratory chain.

Complex III, containing iron-sulfur clusters and cytochromes b1, b2 and c1, catalyzes the oxidation of ubiquinol back to ubiquinone, reducing cytochrome c, located in the inner membrane and intermembrane space. Since cytochromes can take stet-by-step only one-by-one electron, the protons should be released into the intermembrane space.

Complex IV (cytochrome oxidase) has two porphyrin rings with copper and iron atoms. Electrons from cytochrome C fall to iron and copper, and then – to oxygen, which, attaching protons, converses to water molecule.

3.2 NADH Dehydrogenase

The NADH dehydrogenase complex I has two parts – membrane and peripheral [5] (Fig. 3.4). It is believed that in mammals this enzyme consists of 40–45 subunits, some of which are encoded by nuclear DNA, and some – mitochondrial DNA. It contains 6–8 iron-sulfur clusters and non-covalent bound flavin mononucleotide (FMN). But may be in actually the flavin is not FMN but FAD, which is quickly hydrolyzed by the neighbor iron-sulfur cluster to FMN during purification of the enzyme (this suggestion needs in verification).

Oxidizing in the active center, NADH is able to give either (a) two electrons to clusters, or (b) one hydride ion (H^-) to flavin or ubiquinone. Clusters can restore ubiquinone (Q) to ubiquinol Q-2e. Flavin can also restore ubiquinone to ubiquinol in the form of Q-H_2.

Complex I can oxidize the NADH molecule, which was formed in matrix in the cycle of tricarboxylic acids. Electrons from NADH are used to restore ubiquinone Q, which transfers them to cytochrome-bc1-complex III.

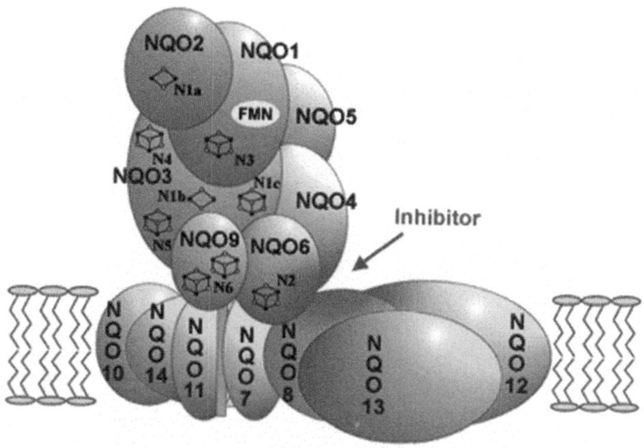

Fig. 3.4 Main subunits of the NADH dehydrogenase complex I

The NADH dehydrogenase complex works as a proton pump. It was assumed that for each oxidized NADH and reduced Q, four protons are pumped through the inner membrane into the intermembrane space:

$$NADH + H^+ + Q + 4H^+{}_{in} \rightarrow NAD + QH_2 + 4H^+{}_{out}$$

But the problem arises: there are practically no free protons in matrix. Therefore, the more realistic reaction scheme, where the proton is taken from NADH:

$$NADH + Q \rightarrow NAD + Q - 2e + H^+{}_{out}$$

Flavin, but not iron-sulfur clusters, can serve as a carrier of protons in NADH dehydrogenase. Cluster of the FeS type takes only an electron from H, but the proton H^+ goes into the intermembrane space or into the matrix.

In the review [5] 8 different schemes of transfer of reducing equivalents in the NADH dehydrogenase complex were described. In some schemes, the electron acceptor from NADH was FMN, in others – iron-sulfur cluster.

According to one of the schemes (Fig. 3.5) of the NADH oxidation, two electrons are first transferred to flavin, and then they are one by one transmitted through a chain of iron-sulfur clusters to ubiquinone and reduced it to Q^{-2}. Cluster N1a, located near the flavin, probably oversees, exchanging the electron with it. If a mobile FMN works here, but not a covalently bound FAD, then it will limit the transfer rate due to the slowness of its "shuttle" diffusion.

Also, a model exists, in which one of electrons from the flavin goes to ubiquinone and the second electron is stored in the cluster N1a and later returned to the main circuit via flavin-semiquinone. It is assumed that this mechanism can reduce the formation of superoxide on the reduced flavin. Against such way of preventing the formation of superoxide, anybody could argue that (a) the formation of superoxide is usually occurs at the level of NADH, and flavin is hidden inside the protein globule, (b) flavin and ubiquinone are too slow carriers due to their difficult diffusion in viscous membrane, (c) the binding site for ubiquinone seems to be located in a

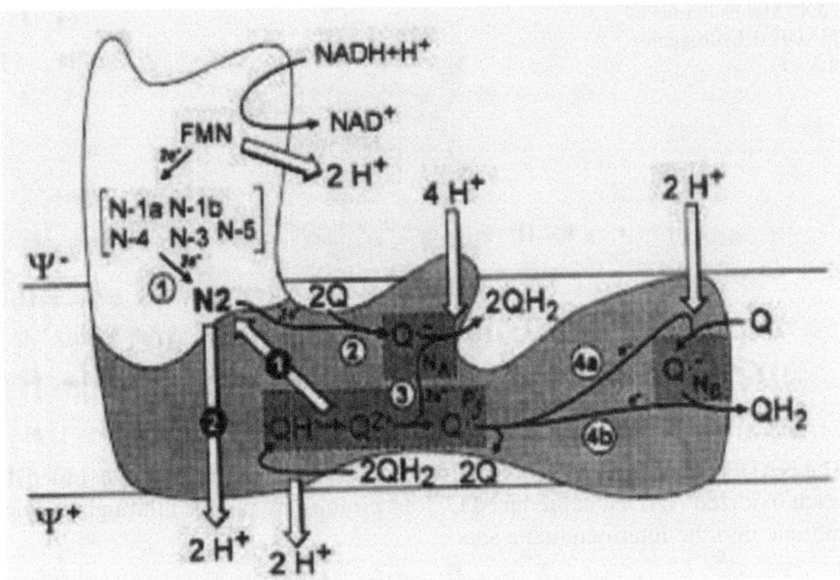

Fig. 3.5 Scheme of electron and proton transfer inside NADH: ubiquinone oxidoreductase

different subunit of the protein, remote from the active site, (d) "drive" the electron back and forth is energetically unprofitable.

The question arises, why do NADH oxidation and ubiquinone reduction require so many protein subunits, as well as several Fe-S clusters? For success oxidation of NADH with reduction of ubiquinone, even 1–2 subunits would be enough. For example, alcohol dehydrogenase (only 70 kDa), consisting of a pair of identical subunits and having neither Fe-S clusters nor flavins, easily converts NADH to NAD, reducing aldehyde to alcohol. Moreover, alcohol dehydrogenase is able to restore ubiquinone upon NADH oxidation [6].

The role of iron-sulfur clusters consists of not so much in the catalysis of the oxidation of NADH molecules, as in the transfer of an electron over a long distance (there is a chain of clusters N3, N1b, N4, N5, N6a, N6b and N2 for this purpose). The chain of clusters (Fig. 3.6) allows easily transfer the electron to a distance of 80 Å. The 2Fe-2S and 4Fe-4S clusters are needed for 2- and 4-electron transport, respectively.

3.3 Succinate Dehydrogenase

Succinate dehydrogenase complex II (~130 kDa) functions in the tricarboxylic acid cycle and in the respiratory chain. It catalyzes the oxidation of succinate (succinic acid salt) to fumarate, reducing electron acceptors, for example, ubiquinone – to ubiquinol:

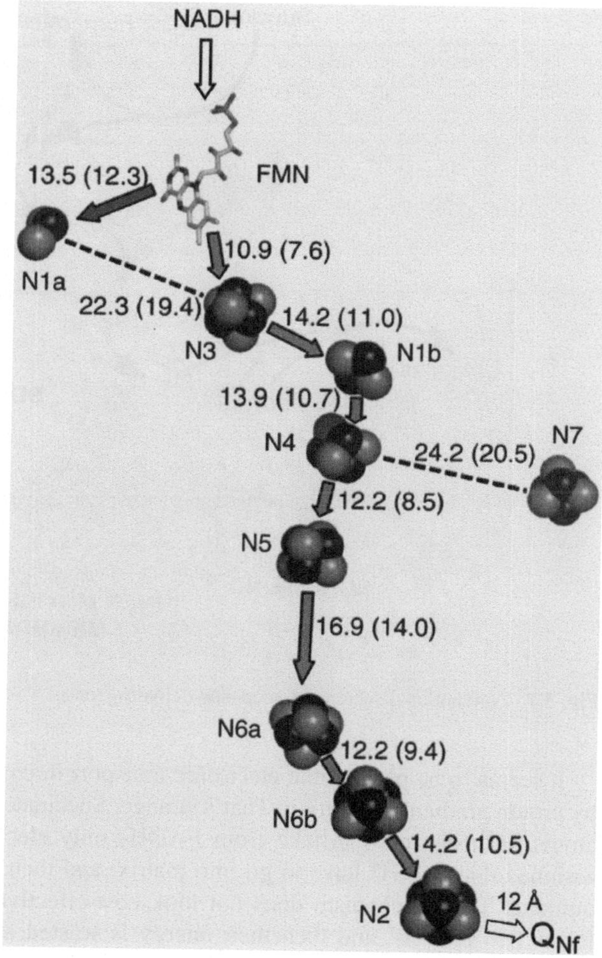

Fig. 3.6 Red-ox centers of NADH dehydrogenase

$$Succinate + Q \rightarrow Fumarate + QH_2$$

Complex II consists of 4 subunits encoded by the nuclear genome. Two hydrophilic subunits are faced to matrix: subunit A is flavoprotein and subunit B is iron-protein. In subunit A there are the succinate site and covalently bound FAD, and in B there are iron-sulfur clusters. The hydrophobic C and D subunits are transmembrane proteins, containing heme b (cytochrome b560) and ubiquinone (Fig. 3.7) [7–13].

Mutations in subunits of succinate dehydrogenase can lead to encephalopathies, tumors and also to the production of superoxide and, consequently, to decrease in life expectancy [14, 15].

The enzyme has a chain of red-ox centers: FAD and three iron-sulfur clusters, transferring electrons from succinate to a distance of 40 Å. The distance between adjacent red-ox centers does not exceed 14 Å [7–13].

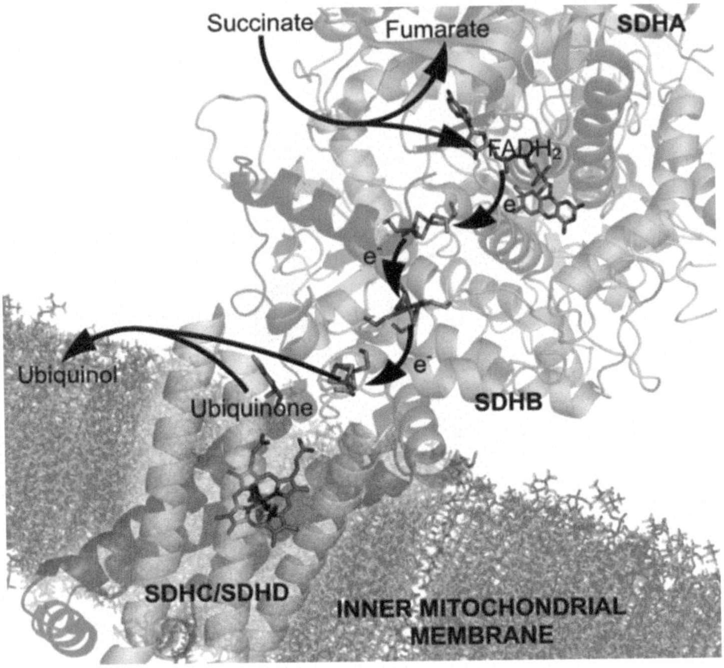

Fig. 3.7 Conventional scheme of succinate dehydrogenase

It seems to be proved that electronic transport in complex II is not accompanied by proton gradient generation. That's strange. Succinate gives to FAD two hydrogen atoms. Iron clusters can take from $FADH_2$ only electrons, but not protons. It is assumed that two H^+ have to go into matrix, and then get out of it when restoring quinone. This mechanism does not look cost-effective: the separation of charges forms two protons, and then their energy is wasted at neutralizing by the matrix buffer.

It is believed that the enzyme has a special binding site for ubiquinone, located in special place between subunits B, C and D. The hydrophobic space between subunits (it places below of heme b) is filled by molecules of cardiolipin and phosphatidylethanolamine. There is an evidence that ubiquinone, taking the first electron is converted to the radical semiquinon, which after receiving a second electron from the cluster 4Fe-4S is restored to ubiquinol [7–13]. This mechanism is a bit doubt, because semiquinon- radical is very active and dangerous. It can induce a free-radical lipid peroxidation.

It is strange, but the function of heme b is still not known. It has been suggested that the first electron, received by ubiquinone, can quickly go to the heme and then quickly return. However, the meaning of such a scheme is not clear.

It would seem that rotenone and piericidin, similar on chemical structure and hydrophobicity with ubiquinone, must inhibit succinate dehydrogenase, but – not inhibit.

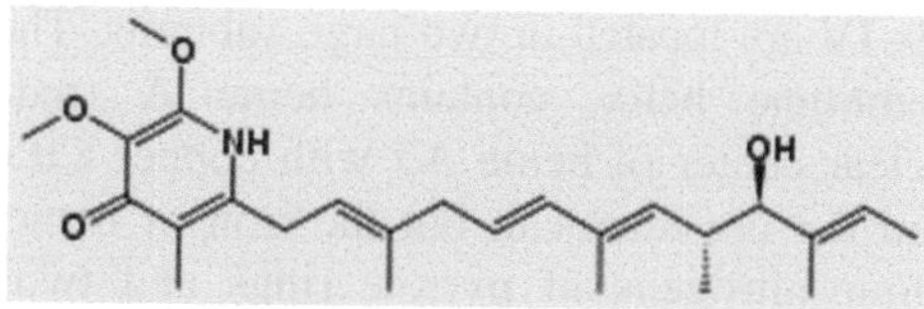

piericidin

It follows that in fact there is no any special site in succinate dehydrogenase – pocket – for binding of ubiquinone. Hydrophobic ubiquinone molecules in the lipid phase simply stick to the protein outside, near heme b. The transfer of electrons to ubiquinone should occur only at the absence of cytochrome C, which easily falls out into the intermembrane space. At the presence of cytochrome C, electrons have to pass directly from heme b to heme c, without ubiquinone.

After all, the electronic d-orbital of these two hems overlap even at a distance of 25 Å between their centers. At the presence of heme b there is no need in ubiquinone, diffusing too slowly. That's why the succinate dehydrogenase is not sensitive to rotenone and piericidin.

Competitive inhibitors-analogues of succinate include: synthetic malonate, as well as components of the Krebs cycle – malate and oxaloacetate.

Although the binding constant of succinate with the enzyme is not large (about $10^3\ M^{-1}$), but the maximal respiration in suspension of MCs is observed with succinate (and not with other di-and tri-carboxylic acids or NADH). This is due not only to that succinate dehydrogenase is directly integrated into respiratory chain, but also to the relatively high permeability of the inner membrane for succinate. It is commonly said that succinate dehydrogenase monopolizes the respiratory chain.

When we are not talking about the MC suspension, but study the MCs in cells, the dominate should not succinate dehydrogenase, but – NADH dehydrogenase, since only one succinate molecule but three NADH molecules in the Krebs cycle exist, and the binding constant of NADH with the dehydrogenase almost three orders of magnitude higher than those for succinate with succinate dehydrogenase. So, NADH in the intracellular MCs has to monopolize the respiratory chain.

3.4 Cytochrome Oxidase

Cytochrome-C-oxidase (cytochrome oxidase, cytochrome Aa3, cytochrome C-oxygen oxidoreductase, complex IV) is the terminal oxidase of the respiratory chain that transfers electrons to molecular oxygen to form water molecule. The mammalian complex IV consists of 13 subunits, of which three large, encoded mitochondrial DNA have catalytic activity, and ten subunits encoded by nuclear DNA have auxiliary functions [16–20].

The membraneous complex IV in MC was detected as a dimer with a weight of 350 kDa. Few monomers have twice as high catalytic activity [21]. It can be assumed that the monomer form is physiological, and the dimeric form is an artifact result of membrane processing during studies.

The cofactors of complex IV are located in two large subunits. The first 12 subunits, forming a transmembrane helix, contains heme A (red-ox potential +0.22 volts) and the bi-nuclear center of heme A3 with copper Cu (+0.24 volts). Chemically the hems A and A3 are identical, but the iron of heme A forms six coordination bonds with four nitrogens of pyrrole rings and two nitrogens of histidine. Iron of heme A3 forms only five bonds, leaving the sixth one for oxygen. There is a strong anti-ferromagnetic interaction between iron and copper of the binuclear center. In the second subunit there is a Cu-Cu-center (+0.7 volts), consisted of two copper atoms [22–25].

It is believed that complex IV consistently oxidizes 4 cytochrome C molecules and, taking 4 electrons, reduces O_2 to $2H_2O$. In addition to cytochrome C, direct electron donors for cytochrome oxidase can also serve ascorbate and ubiquinol, but this is often forgotten.

It is assumed that during O_2 reduction, four H^+ from the matrix are captured to form two H_2O molecules, and four H+ are actively pumped into the intermembrane space, creating a proton gradient [23–28]:

$$4\ cit.c^{2+} + O_2 + 8H^+{}_{in} \rightarrow 4\ cit.c^{3+} + 2H_2O + 4H^+{}_{out}$$

The equation is highly conditional, since the stoichiometry for the transmembrane proton has not been proven. Moreover, all protons, including "substrate's protons", originating from NADH or succinate, are allegedly taken from the matrix buffer, but there are practically no free protons (when at least one H^+ appears inside one MC, it immediately is captured by the matrix buffer). The exact mechanism of proton transport is still unclear. It is not clear also how couple the pairing energy of the electrons with the movement of protons.

It is assumed that complex IV has a channel that connects the aqueous phase of matrix with the binuclear center and serves to deliver "substrate" protons for the formation water from them and oxygen. Another channel, apparently, forms a transmembrane end-to-end path [24–28].

Cytochrome C binds to the second subunit and restores Cu-center, located near the surface of the membrane. From this center, the electron goes to the heme A and the binuclear center further to A3-Cu in the thickness of the membrane. In the binuclear centre, the binding of O_2 and restoration to H_2O takes place [23–28]. Oxygen in the process of recovery to water releases a large amount of free energy.

Oxygen in the binuclear center is not in the peroxide form, as previously believed, but in the oxo-ferrilic form, where the bond between the oxygen atoms is already broken. Reduction of oxygen occurs by the pair-wise transfer of electrons, which allow eliminate the formation of dangerous superoxide. A fully restored bi-nuclear center binds O_2 and then the formation of oxygen adduct takes place. Four electrons are transferred to oxygen: two are supplied by the iron of heme A3, one by copper, and the fourth by tyrosine, which also gives a proton to break the bond in O_2 (Fig. 3.8). The resulting neutral tyrosine radical is reduced to anion by cytochrome.

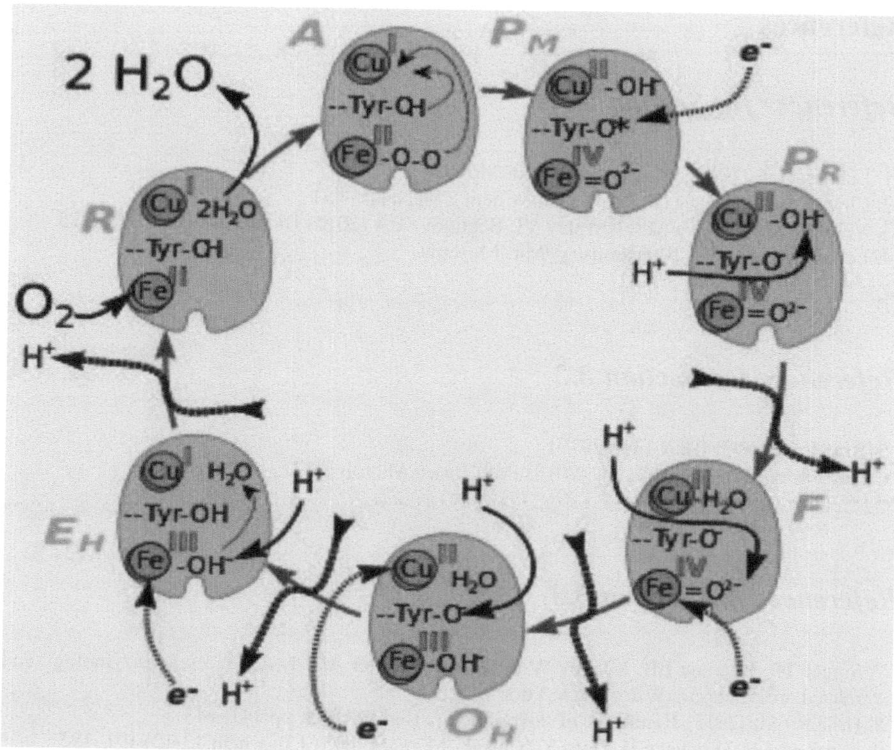

Fig. 3.8 Scheme of molecular oxygen reduction to water in the active center of cytochrome oxidase

Protonation of Cu occurs with the formation of a water molecule, which is bound by copper coordination. Iron in FeO_2 is recovered, and the bound oxygen is protonated. The first molecule of water is released. A tyrosine anion get a proton, and the Cu is reduced by cytochrome C. Iron is recovered, after which the associated OH-group is protonated to form a second water molecule. Now the binuclear center is fully restored again and ready to bind a new oxygen molecule.

Eighteen positions for phosphorylation were found in the subunits of complex IV, but the role of phosphorylation still was not determined [29]. And there is there the synthesis of ATP, as ATP-synthase? It is no accident that the functioning of cytochrome oxidase inhibited by excess ATP…

Cyanides, sulfides, azides, carbon monoxide and nitric oxide bind to the binuclear center and, competing with oxygen, inhibit the enzyme, which leads to cell death from asphyxia. Mutations in cytochrome oxidase lead to serious metabolic disorders in organs with high energy consumption (brain, heart, muscles). Diseases associated with dysfunction or impaired cytochrome oxidase assembly is the most severe [30].

References

References for Section 3.1

1. Lehninger A (1966) Mitochondria. Mir, Moscow
2. Chance B, Hollunger G (1963) J Biol Chem 278(1):418–431
3. Vekshin NL, Frolova MS, Kovalev VI, Begunova EA (2015) Biophysics 60(1):129–135
4. Lehninger A (1974) Biochemistry. Mir, Moscow

References for Section 3.2

5. Brandt U (1997) BBA 1318:79–91
6. Kamenskyh KA, Vekshin NL (2018) Prikl Bioch Microb 54(3):316–319

References for Section 3.3

7. Singer TP, Kearney EB, Kenney W (1973) In: Meister A (ed) Advances in enzymology and related areas, vol 37. Wiley, New York
8. Hille R (ed) (2013) Handbook of flavoproteins. Berlin, vol 2, pp 141–143
9. Tomitsuka E, Hirawake H, Goto Y, Taiwaki M et al (2003) J Biochem 134(2):191–195
10. Yankovskaya V, Horsefield R, Törnroth S et al (2003) Science 299(5607):700–704
11. Sun F, Huo X, Zhai Y, Wang A, Xu J, Su D et al (2005) Cell 121(7):1043–1057
12. Horsefield R, Yankovskaya V, Sexton G et al (2006) J Biol Chem 281(11):7309–7316
13. Iverson TM (2013) BBA 1827(5):648–657
14. Bardella C, Pollard PJ, Tomlinson I (2011) BBA-Bioenergetics 1807(11):1432–1443
15. Yang M, Pollard PJ (2013) Cancer Cell 23(6):709–711

References for Section 3.4

16. Michel H (2013) Structure and mechanism of Otto Warburg's respiratory enzyme, the cytochrome c oxidase
17. Pierron D et al (2012) BBA-Bioenergetics 1817(4):590
18. Sotoa IC et al (2012) BBA-Bioenergetics 1817(6):883
19. Taanman JW (1997) J Bioenerg Biomembr 29(2):151–163
20. Kadenbacha B et al (2015) Mitochondrion 24:64–76
21. Sone N, Takagi T (1990) BBA 1020(2):207–212
22. Tsukihara T et al (1995) Science 269(5227):1069–1074
23. Yoshikawa S, Shimada A (2015) Chem Rev 115(4):1936–1989
24. Wikstrom M (2012) BBA-Bioenergetics 1817(4):468–475

25. Gorbikova EA, Belevich I, Wikstrom M, Verkhovsky MI (2008) PNAS 105(31):10733–10737
26. Sharmaa V et al (2015) PNAS 112(7):2040–2045
27. Fadda E et al (2008) BBA 1777(3):277–284
28. Konstantinov AA (2012) FEBS Lett 586(5):630–639
29. Arnold S et al (1998) Eur J Biochem 252(2):325–330
30. Houstkova P et al (2004) Physiol Res 53(Sup. 1):S213

Chapter 4
Physico-Chemical Mechanisms

4.1 Interactions in Enzyme-Substrate Complexes

Connecting with the enzyme, the substrate (succinate, NADH, etc.) is largely purified from the water coat and is located in the space of the desired way. Moreover, the enzyme limits the number of degrees of freedom, directing the reaction along a certain spatial coordinate.

The main role in intermolecular interactions is played by external electronic orbital. Electronic transitions are instantaneous and the cores are then slowly reconfigured. The energy contribution of nuclear vibronic movements is an order of magnitude less than that of electronic ones. Electron-vibrational transitions are quantized, not continuous, as they traditionally depicted [1, 2] by a trajectory in the coordinates "energy / inter-nuclear distance".

Joining the substrate to the enzyme is accompanied by neutralization (partial or complete) of opposite charges, as well as the formation of ionic and hydrogen bonds. Sorption of the substrate on the enzyme leads to releasing a large quantum of chemisorption energy E_{chem} [1]. This quantum is used for almost all non-activated leakage further stages. When complete neutralization of opposite electrical charges takes place, the magnitude of E_{chem} can be found according to the equation:

$$E_{chem} = 2,15 \times 10^6 N q_1 q_2 / \varepsilon r$$

Here q_1 and q_2 are the magnitude of positive and negative charge, N is the Avogadro number, r is the final distance between neutralities charges ε is dielectric constant. At r = 1.5 Å (length of covalent bond) and ε = 1 we get the maximum E_{hem} = 250 kcal/mol. For enzymes you can take ε~1, because water molecules (ε = 81) are not usually placed between neutralities groups. With incomplete neutralization of charges (dipoles, etc.) or at ε > 1, the value E_{chem} is much smaller, but still it is higher than the activation energy.

N. Vekshin, *Biophysics of Mitochondria*,
https://doi.org/10.1007/978-3-030-33853-4_4

Not all E_{chem} is spent for overcoming of an activation barrier and commission of useful work because a small part is lost as heat. Sorption heat ΔH, measured calorimetrically, is a small lost energy: $\Delta H = E_{chem} - E*$, where $E*$ is the energy of the transition electron-vibrational excited complex (EVEC). To describe the energy profile of enzymatic reactions useful idea that in the active center of the enzyme a mixing of molecular orbitals of the ground and excited state of the substrate takes place, that leads to changing of geometry, dipole moment and pK of the substrate [2].

The loss of a small part of the energy as the heat ΔH during the sorption of the substrate (S) on the enzyme (En) provides a local irreversibility of the further stages. In this case, the reversibility of the enzymatic reaction, as such, is preserved: the direct and reverse pathways coincide: S + En = EVEC of S-En = Pr + En. The reversibility is due to the fact that ΔH usually does not exceed several kcal/mol, and therefore ΔG can easily be changed up to the change of sign due to the entropic term TΔS [3]. By the way, the reverse transfer of electrons from succinate (through succinate dehydrogenase) to NADH dehydrogenase in the membrane is as non-energy-consuming as the reduction of NAD with ethanol by alcohol dehydrogenase in solution. The resulting EVEC S-En is quickly rebuilt and then transformed with the release of the product (Pr). The energy of EVEC, eventually, allows the flow of stages of the transfer of electrons and protons, and desorption of Pr.

All of the above said is fully applies to all substrate-enzyme complexes. For example, when NADH is bound in the active center of NADH dehydrogenase, the formation of an EVEC complex occurs, in which the substrate is excited to a state with a large dipole moment and sharply changed pK of the pyridine ring, which leads to easy cleavage of the proton and two electrons.

4.2 Mechanisms of Electron Transport

Transfer of electrons and protons in the respiratory chain is carried out with the participation of red-ox centers formed with iron-sulfur clusters (Fe-S, Fe_2-S_2, Fe_4-S_4), flavins (Fig. 4.1), cytochromes (B,C,A), and ubiquinone. The proton is heavier

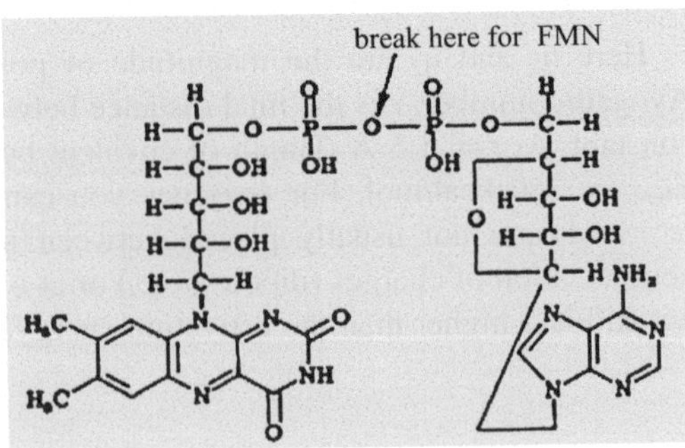

Fig. 4.1 Structural formula of FAD

than electron in the 1840 time. That is why it moves so much slower. Electron transport is separated from proton transport to get a maximum speed and to get the transmembrane potential.

Flavin and iron-sulfur clusters themselves (without enzymes) can easily receive and give an electron. However, they are hidden deep into protein globules. Polypeptide chains are good dielectrics. Somebody talk about semiconductor properties of peptide globules [4], but such point of view is purely speculative. Therefore, it would seem, the electron should not be transferred through a protein matrix over long distances (10–15 Å). But it does. Why?

Because in fact, electrons do not jump between the centers (nuclei) of atoms, but «flow» between overlapping electron clouds, stretching up to 5–15 Å [5]. Conventional drawings of red-ox centers without electronic clouds create the illusion of long distances and are often misleading us.

In hetero-aromatic molecules, the average distance from nucleus to the maximal electron density is only 1–2 Å. The probability of detecting an electron at a distance from the nucleus decreases exponentially, for sp-orbitals – almost to zero at 5 Å [5, 6]. Figure 4.2 shows the distribution of the electron density in the pyridine molecule, which is an analogue of the pyridine ring of NADH.

In the case of flavins and NADH, there is a probability of finding an electron at a distance of 5 Å [6]. In the electronically excited or restored state of flavin and NADH, their electron clouds are delocalized even more [6]. Figure 4.3 shows the radial distribution of the electron density for the excited (reduced) states of different orbitals. For 3 s-, 3d – and 3p-orbitals, the electron density at 10 Å is high and drops to zero at 15 Å [6, 7].

The electron clouds of iron ions occupying the d-orbitals extend to ~ 10–15 Å [7]. Electron clouds of iron-sulfur clusters and cytochromes, penetrating the thickness of the protein globule, reach the surface and therefore can contact each other. That provides a fast transfer of the electron.

No special "electronic trails" through protein globules are required. However, hetero-aromatic amino acid residues (tryptophan, tyrosine, phenylalanine) can influence on the electron clouds of red-ox centers by their own electron clouds.

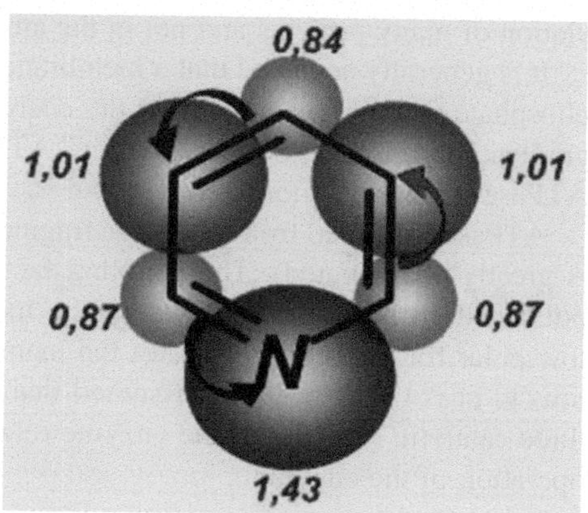

Fig. 4.2 Electron density distribution in the pyridine molecule (digits are fractions of the electron density)

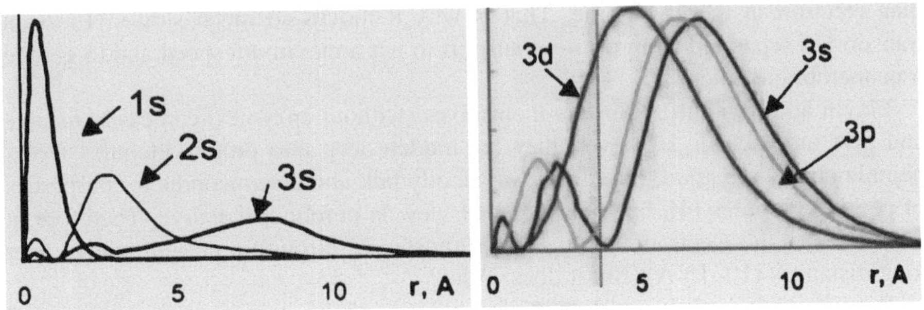

Fig. 4.3 Radial distribution of electron density for various orbitals

4.3 Membrane's Synthesis of ATP

One of the main functions of MC is the synthesis of ATP — a convenient molecular form of chemical energy storage. It is convenient because it is used in a variety of intracellular processes for their activation by phosphorylation of proteins.

ATP can be formed from ADP and phosphate as a result of substrate phosphorylation in the liquid phase (in glycolysis) and membrane phosphorylation in the respiratory chain. MCs implement both ways, the first of which occurs in matrix, and the second one is associated with crystae [8].

The first stages of the aerobic pathway (up to the stage of formation of pyruvic acid) coincide with anaerobic glycolysis, but in the presence of oxygen the pyruvic acid is not converted into lactic acid, and enters into the Krebs cycle. One molecule of pyruvic acid gives 15 "macroergic" ATP molecules.

In fact, there is no special "macro-energeticity", which many biochemists often say concerning to ATP molecule. Not so much energy is released during the hydrolysis of ATP to ADP. It is not more 7 kcal/mol. But ATP can easily phosphorylate many proteins that lead to activating of them. Therefore, ATP is the universal "energy currency" of a cell. For example, the hydrolysis of ATP at the actomyosin complex reduces muscle fibers. The "macroergicity" of ATP is the ease phosphorylation of many proteins and not in the amount of energy as such.

It is generally accepted that a membrane ATPase (500 kDa) catalyzes the ADP – phosphate reaction by converting the energy of the transmembrane electrochemical gradient of protons into a "macroergic" ATP molecule [8]. It is believed also that the ATPase strongly conforms.

ATPase is formed from two large fragments (Fig. 4.4; the length of the lipid phase is greatly exaggerated). The coupling factor F1, consisting of 9 subunits, is quite polar and strongly protrudes from the membrane thickness into the matrix. The low-polar fragment F0 permeates the membrane through and has channels for protons to pass through. It was assumed that when protons pass through the channel, three catalytic subunits of the enzyme rotate and allegedly this is what ensures the operation of the enzyme.

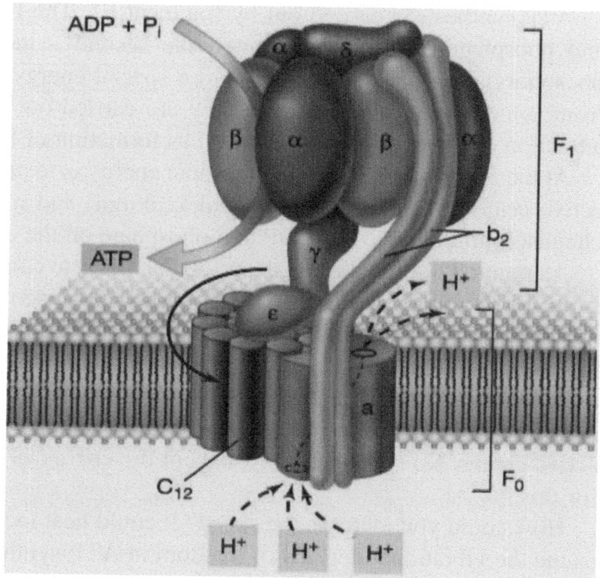

Fig. 4.4 The structure of ATP synthetase

ATP synthetase, like any other normal enzyme, and is able to lead the reaction towards the synthesis of ATP, and in the direction of hydrolysis:

$$ADP^{3-} + HPO_4{}^{2-} + H^+ \leftrightarrow ATP^{4-} + H_2O \qquad \Delta G = -7\text{kcal/mol}$$

At pH = 7.5, when phosphate groups are ionized, the synthesis reaction requires a proton, and the hydrolysis reaction does not. Moreover, the proton does not necessarily have to be transmembrane. Any proton donor can catalyze ATP synthesis. That is why the ATP synthesis was observed at acid attack on pre-broken membranes [9].

In the reaction of ATP synthesis, the number of charged groups decreases (from five negative and one positive, four negative groups are obtained). In fact, there is a neutralization of charges with the release of energy: one minus neutralizes one plus, which is accompanied by the accumulation of energy in ATP.

Almost twice much energy can be obtained by neutralizing the charges H^+ and OH^-

$$H^+ + OH^- \rightarrow H_2O \qquad \Delta G = -13.7\ kcal/mol$$

Such reaction in the cell could give energy for synthesis of two ATP molecules.

As well-know, upon draining in a test tube together a strong acid and alkali the energy spend to heat so intensively that the solution is noticeably warmed. If the neutralization of a proton with a hydroxyl-anion occurs not in solution, but in the lipid or protein phase, then part of the energy may go not to heat, but to serve for a phase transition or conformational change.

ATP synthesis is carried out by fragment F1. The first stage is binding of ADP and phosphate in the active center; the second – their protonation; the third – phosphorylation of ADP to ATP. No a special energy for these stages is required from somebody outside, because they are carried out spontaneously – due to the attraction of opposite charges, as well as formation of hydrogen bonds.

At the fourth stage, it seems that some energy is required to release ATP from the active center, but it is small (several kcal/mol) and may well be drawn from the chemisorption energies of ADP and phosphate on the enzyme or – from heat tank.

As pointed out by N.M. Bazhin (2017), the rate constant for desorption of ATP from ATP-synthase is $k_{-ATP} = 10^{13} \cdot \exp(-60/2,48) = 300\ s^{-1}$ (here 60 kJ/mole is activation energy), which is sufficient for some production of ATP without the involvement of additional outside energy. In other words, the thermal energy of the medium is enough for a small rate of desorption. In place of the desorbed ATP molecule, another ATP molecule will immediately come from the neighboring active center. No special energy, except the energy of thermal motion, is required for this.

How could you increase the speed? It could heat locally the active center, i.e., to excite the vibrational degrees of freedom in ATP-synthase where sits ATP [10].

Desorption of the product is the limiting factor of many enzymatic reactions. Membrane ATPase is no exception. It is necessary that the active center will receive enough energy for ATP desorption [10]. But this energy is small, about 1 kcal/mol. If we assume, as some biochemists, that the desorption ATP from the enzyme requires as much as 7 kcal/mol, the usual average thermal energy of 0.7 kcal/mol (at 37 °C) is clearly not enough. And, having spent 7 kcal/mol by any "third-party" method for desorption, we will not win anything in the balance: how much we got in the synthesis of ATP, the same amount spent on desorption.

If the inner membrane of MC has a channel for the output of ATP molecules (and the inlet channel for ADP), a special "energy" from somebody is also not necessary. Just the membrane ensures that desorption is not energy-consuming. If the ATP molecule is released from the active center due to the local increased thermal energy (0.7 kcal is the average thermal energy), then it immediately goes out along the concentration gradient, because the local ATP concentration in matrix is high, and outside it is almost zero. The outgoing ATP is exchanged for the incoming ADP.

4.4 Proton Transport and Chemiosmotic Hypothesis

In 1961 (long before Boyer's conformational hypothesis), Mitchell [11] put forward the idea that proton transport (associated with electron carrier along the respiratory chain) through the inner membrane into the intermembrane space creates a transmembrane potential, which is necessary for the synthesis of ATP.

In the original version, this hypothesis had the advantage over chemical hypothesis in the sense that it does not contain "intermediate macro-energetic" X~Y. In the second version [12] Mitchell resorted to X~Y. Subsequently he refused it.

The existence of some potential on phosphorylating membranes was established experimentally [13–16], which was perceived by many as a triumph of the Mitchell scheme.

However, it was not so simple. The chemiosmotic model was not only sharply criticized (in particular, for what is essentially thermodynamic and, therefore, cannot explain anything in kinetic mechanisms) [17, 18], but came into conflict with many facts. For example, Maloff et al. [19], using microelectrode technique on a huge MC (5-10 micron) from the liver of fed cuprizone mouse received the membrane potential not −200 mV, but only 16 mV, and of opposite sign. Komai et al. [20] observed partial conservation of the ATP-phosphate exchange and phosphorylation oxidation of NADH in sub-mitochondrial non-vesicular particles. Knowles and Racker [21] and Vale et al. [22], found the synthesis of ATP on purified calcium ATPase of sarcoplasmic reticulum and concluded that the energy source is the binding of calcium to protein. All mentioned authors made a conclusion about the inaccuracy of Mitchell's scheme.

Yaguzhinsky et al. [23] using the potential-sensitive dye Dis-C3-5 and the distribution of tetraphenylphosphonium, found that the oxidation of succinate in suspension of MCs gives the potential of −170 mV, but the addition of ADP reduces it by only 3–6 mV, and malonate or cyanide sharply reduced the rate of respiration without affecting the potential. The addition of ammonium acetate, which removes the gradient of transmembrane protons, did not change the effect of inhibitors on respiration and potential. The obtained results were talking about the uselessness of transmembrane potential for coupling with ATP synthesis.

One of the decisive experiments, which seemed to confirm the chemiosmotic hypothesis, was experiments on acid-alkali shock of MCs [24] and chloroplasts [25]. However, these experiments of Mitchell et al. [24] and Jagendorf [25], seems to be showing a need for transmembrane proton gradient for ATP synthesis, was not as probative as first thought. In particular, it was found that it is possible to observe the emergence of ATP in MCs and chloroplasts, when the sign of the gradient of protons is opposite to the requirements of the chemiosmotic hypothesis, and in magnitude considerably less than required [26, 27]. In the isolated MCs, inactivated by artificial "aging", the ability to appear ATP at pH-shock remained. It was noted that the appearance of ATP does not depend on the pH value, but on the final pH [17, 26, 27].

In 1974, the beautiful experience of Racker and Stoeckenius [28] on the chimera, created from bacteriorhodopsin, mitochondrial ATPase and phospholipids, became, in the figurative expression of V.P. Skulachev, "the weight that tipped the scales": when bacteriorhodopsin was illuminated, the pH value was changed and ATP was synthesized. It would seem that the "chimera" worked in full accordance with Mitchell's predictions.

But after a few years, "weight" began to outweigh the other side. Thus, in experiments with bacterial photophosphorylation, a discrepancy between the pH gradient across the membrane and ATP synthesis was observed [29]. At pH = 6 the synthesis of ATP in upon illumination was increased three times without any change in gradient of pH and of the potential [30]. And so on.

The analysis of various data on the above-discussed subject was made in [18]. One of the most important conclusions was that no "special" energy is needed for the act of ATP synthesis. Energy is needed only to remove ATP from the active center, and it is not great.

In the original Mitchell's hypothesis it was assumed that protons are taken from the dehydrated substrate [21] (and this is correct). Subsequently, under the pressure of "clarifying" experiments, Mitchell and his followers mistakenly claimed that protons are taken from the aqueous phase of matrix.

The question "to backfill": how many free protons contained in a single spherical MC with volume V and diameter d = 1 micron, if at least half of its volume is occupied by crystae, ribosomes and DNA, and the pH of the matrix is 8?

Answer: ~ 0. Compare the answer to the similar task on page 485 and 911 in the book of Lehninger [31], taking $V = 4/3\ \pi\ (d/2)^3$. There are almost no free protons in aqueous phase of matrix. If suddenly appears at least one H^+, it immediately disappears, grasping the matrix buffer. Various chemiosmotic schemes with taking from 2 to 8 free protons from aqueous phase of matrix are purely speculative. The protons of matrix are connected by phosphates, proteins, etc. All sorts of newfangled "measuring" the pH inside of one MC are incorrect, because the pH-sensitive dye itself dramatically knocks down the pH inside the matrix. However, the intramatrix pH found by pH-metrically latching the medium in the suspension of burst MCs is indeed equal to 8, this is a statistically correct value.

The real sources of protons are dehydrated substrates (NADH, succinate, etc.), as well as phosphate buffer, available in matrix. Protons of substrates are immediately ejected into the intermembrane space, bypassing the aqueous phase of matrix. A high local concentration of protons is created in the inter-membrane space, since its width does not exceed 200 Å. Any H^+ can penetrate into the channel of membrane-bound ATP synthase and catalyze the synthesis of ATP from ADP and phosphate.

According to the original Mitchell's scheme, at the beginning of the respiratory chain (at the level of NADH-dehydrogenase and succinate-dehydrogenase), the protons taken from the substrate are ejected into the intermembrane space, and at the end of the chain (at the level of cytochrome oxidase) – are taken from there. According to the Williams scheme, the conjugation is local, i.e. the proton is taken by ATP synthetase at the same point where it was ejected by one red-ox center. A discussion of the arguments "for" and "against" the schemes of Mitchell, Williams, etc. is contained in [32].

4.5 On Measurements of Transmembrane Potential

It is believed that the transmembrane potential is the driving force in the synthesis of ATP [33]. However, before we can say that energy is stored in the form of transmembrane potential and then this potential is used to perform some useful work, it is necessary to measure it reliably. Surprisingly, the question of the correct measurement of transmembrane potential still remains open. It would be pointless to dispute the existence of transmembrane potential, but evidence of its magnitude and significance could be challenged.

One of the most common approaches to the measurement of transmembrane potential is the use of a special electrode sensitive to tetraphenylphosphonium (TPP) (see, e.g., [34] and references in it). This cation penetrates quite well into MC, which are known to be negatively charged both outside and inside (positive charge – in intermembrane space). Then TPP is able to exit from MC, which has been interpreted as a result of reduced potential.

In MC, there are many positively and negatively charged proteins, lipids, DNA, phosphates, etc. Various substances and substrates (succinate, ADP, etc.) added to the isolated MCs also tend to have a charge. Not surprisingly, these supplements affect the redistribution of TPP. Obviously, TPP not so much "measures" transmembrane potential, but reflects the redistribution of charged substances. A strange "potential generation" by exogenous NADH [34] oxidizing supposedly in the outer membrane can be explained from this point of view.

Most often transmembrane potential is determined using spectrally reacting charged dyes [35, 36]. Detecting changes in the optical density of any dye at the absorption band with varying concentrations of potassium chloride, one can calibrate the transmembrane potential in the system under study; see, for example, the use of oxanol-IV in sub-mitochondrial particles [35]. At the same time, it is assumed that all changes in optical density observed with the addition of succinate or other substrates are due to the change in potential. This assumption is too bold.

Since fluorimetry is hundreds of times more sensitive than photometry [37, 38], fluorescent dyes (probes) are used most often. These include rhodamine 123, tetramethylrhodamine, mitotracker red, safranin O, merocyanine 540, Dis-C3-(5), ANS etc.

It is believed that these fluorescent dyes are specific potential-dependent and that they reliably measure the transmembrane potential of both isolated MCs and MCs in cells [34, 38]. But let us turn not to opinions but to facts.

Rhodamine 123

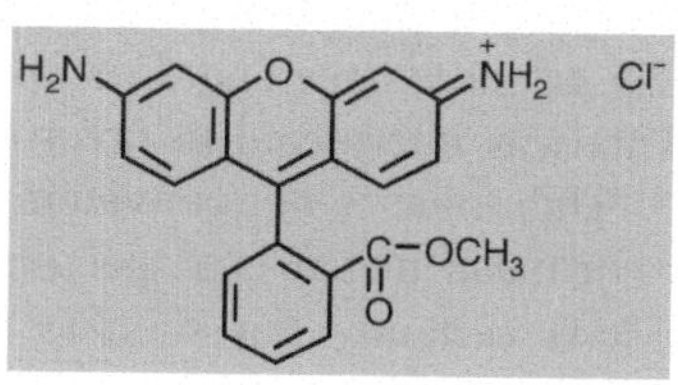

The first publication using rhodamine 123 as the vital dye for MCs appeared in 1980 [39]. Rhodamine 123 is an amphiphilic positively charged molecule. Due to the charge and the large dipole moment, rhodamine 123 should be, in principle, potential-sensitive. The authors of the work [39] found, using a luminescent microscope, that this dye detects a set of thin long (up to 50 micron) curved filaments inside cells. These threads were identified like the MCs. In subsequent publications of these and other authors [40, 41] the thesis of the specific coloring of MCs with rhodamine 123 had not been doubt. But this is despite the fact that the data of electron microscopy and dark-field (native, without dyes and fixators) optical microscopy suggest that MCs in somatic animal cells are usually spherical or ellipsoid structures of the order of 1 micron [42–44].

What were the arguments cited by the authors [39, 40] in favor of the rhodamine-colored threads are really MCs? The arguments are here: (1) Microscopy in the phase-contrast image of individual colored cells revealed exactly the same filamentous structures as in the luminescent mode. (2) Organelles stained with rhodamine 123 swelled and lost the dye when valinomycin was added. (3) Rhodamine 123 did not stain the cytoplasmic and nuclear membranes and apparently did not stain the membranes with lysosomes, endoplasmic reticulum and Golgi apparatus. (4) When dinitophenol or FCCP (the both are uncouplers) was added to the cells, the fluorescence intensity of rhodamine 123 decreased. These were all "proofs".

Why is the word "proof" quoted? Because, firstly, although the structures, visible by phase contrast and luminescent were identical, it does not follow that they were exactly MCs. In order to be sure that they are really MCs, it would necessary to paint them with ethidium bromide or other dye that detects the presence of DNA (DNA is localized in the MC and nucleus). Or somebody could register the succinate: tetrazole-reductase reaction of succinate dehydrogenase inherent only to MCs [44]. Secondly, the swelling of structures is a common property of intracellular organelles, not only MCs. Third, the fact that rhodamine 123 does not stain the cytoplasmic and nuclear membranes is not imply that it stains only MCs. Fourth, there is no evidence in [39, 40], that the dye does not stain the lysosome, endoplasmic reticulum and Golgi apparatus, since general reasoning on this subject cannot be considered as proof.

In [40, 41] it was found that rhodamine 123 is not too specific to MCs; this dye is capable of very intense staining also of myofibrils of cells. Interestingly, colchicine has eliminated the staining of cells with rhodamine 123.

It was shown that rhodamines (3B, 6G, 123) and cyanide dyes in a living cell almost equally stain the same structures [40]. Rhodamine 6G, which is an analogue of rhodamine 123 (it is also positively charged and stained filamentous structures), inhibited oxidative phosphorylation and transport of adenine nucleotides [45], whereas rhodamine B (which has no charge and does not stain) did not exert these effects on isolated MCs. This was interpreted in favor of that rhodamine 123, as rhodamine 6G, stains MCs. This logic is not convincing. If any dye inhibits any functions of MCs, it does not mean that it is a specific marker. In addition, if it inhibits, then it kills, i.e. it is certainly not suitable for vital measurement of transmembrane potential.

In the monograph [44] the situation with rhodamine 123 was critically considered from all points of view. It was emphasized that unlike intracellular MCs, the suspension of isolated MCs is poorly stained with rhodamine 123. And there was done a clear conclusion: rhodamine 123 is not a specific dye for MCs. In living cells, there is a lot of fibrillar actin, which is tightly bound to many MCs and regulates their mobility [46]. It has been shown [46] that rhodamine 123 binds well with fibrillar actin, which is known to have the form of filaments…

Tetramethylrhodamine

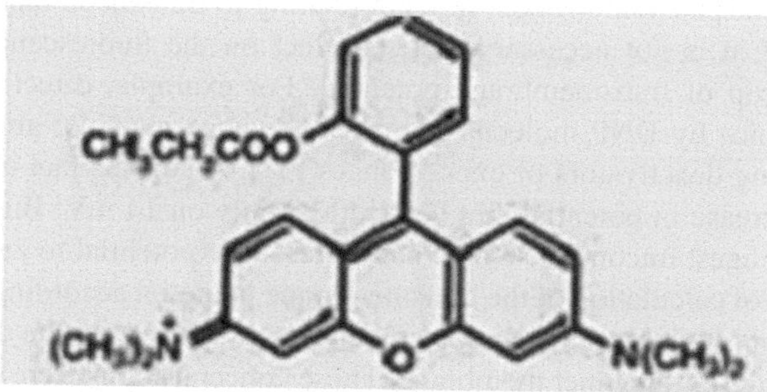

Let consider the data from [47–49]. Here, 4 fluorescent dyes (ethyl ether of tetramethylrhodamine – TMR, safranin O, rhodamine 123 and mitotracker red) were applied to study MCs in not-energized state – at the absence of substrates, energized state – after adding succinate, and de-energized state – after adding DNP, uncoupler. In these works, the fluorescence correlation spectroscopy method, which allows detect individual particles, was used. It was found, in particular, that the amplitude of fluorescence fluctuations of MCs, stained with TMR, varies significantly depending on their energetic state. Hence, it was argued that TMR measures the value of membrane potential at the level of a single MC. In favor of this, it was given the fact that the addition of succinate caused an increase in the amplitude of the fluctuation peaks, which was interpreted by the energy-dependent input of TMR molecules into MC in response to the generation of the membrane potential. The addition of DNP led to a decrease in the number and amplitude of the peaks. This was explained by the release of the dye from the MC during their transition to the de-energized state.

The graphs given in [47, 49] do not cause objections. But the interpretation is not unambiguous. Is the increase in fluorescence intensity and amplitude of fluctuations necessarily caused by the entry of the dye inside the matrix? Isn't there another reason? For example, it is possible to redistribute the dye from the surface of the outer membrane to lipid bilayer or to protein pockets of the same membrane or to the inner membrane (not to matrix). After all, the dye has a higher affinity for the membrane than for water. In particular, TMR has 4 additional hydrophobic methyl groups that increase affinity to membranes. The exit of the dye from membrane into matrix is energetically not favorable. But even if we assume that the dye got into matrix, is its entrance there necessarily caused transmembrane potential? Quite a bit of "fluffing" the membrane and the dye goes outside from the inside just along own gradient. Many membranotropic substances can act as a "leavening agent". Besides, addition of succinate activates the respiration that could lead to decrease in the oxygen concentration in the suspension that should increase the dye fluorescence due to decrease in fluorescent quenching by oxygen. Such alternatives were not even discussed in [47–49].

Succinate is a di-anion that penetrates quite well into MC. Its input can activate the input of the TMR cation via the mechanism of conventional symport of anions and cations. Therefore, it is not surprising that succinate could affect the fluorescence of TMR in the suspension of MCs. It is not relating to transmembrane potential.

As for DNP, it is not necessary that its effect on the fluorescence of TMR is caused by a drop of transmembrane potential. For example, direct quenching of TMR fluorescence by DNP molecules is possible, since phenolic groups and nitro groups are strong deactivators of excited states [37]. A curious fact in [49]: adding DNP led to decrease in potential, but very little – only on 14 mV. But in reality the DNP is the strongest uncoupler, which had to reset the potential to zero. . .

The method of calculation of the transmembrane potential according to the Nernst equation is strongly doubt. Indeed, the equation was applied for the dye concentrations on both sides of the inner membrane. These concentrations were determined by fluorescence. But fluorescence intensity is not always proportional to concentration. For example, in the event of dimers or aggregates of the dye the fluorescence usually decreases. In this case, the frequency of fluctuations on the confocal microscope will be reduced (by increasing the size of particles). Moreover, the concepts of "outside" and "inside" in relation to the dye, which is embedded into the membrane, are not correct. Most dyes have a high affinity to non-polar regions of membranes and fluoresce well from the un-polar or polar membrane phase, not from water.

The calculation of TMR concentration inside a single MC was made in [49] on the assumption that the volume of one MC is 0.17 μ^3. This value is greatly underestimated, because the average diameter of the hepatic MC is about 1 μ. So, the concentration of the dye "inside" was overestimated of 3 times. The potential value of −180 mV calculated in [49] (and not only in [49]) by the Nernst equation is poorly grounded.

Safranin O

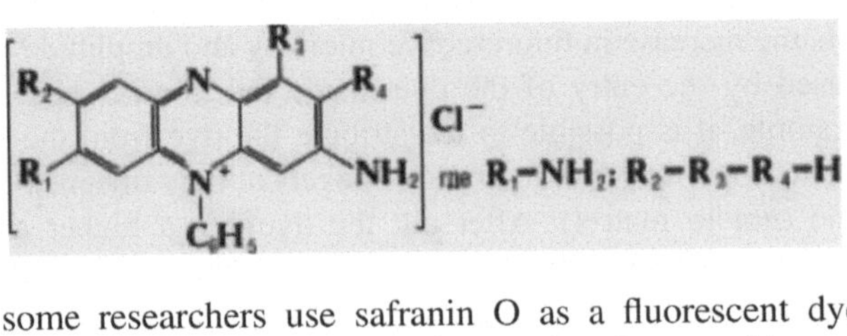

Although some researchers use safranin O as a fluorescent dye sensitive to transmembrane potential, however, there is little direct evidence to prove its suitability.

For example, works [48, 49] provide the data that the safranin O fluorescence was reduced in MCs in response to the addition of succinate that was interpreted as an increase of transmembrane potential. However, first, safranin is not specific to MC; it is known to be widely used for coloring of nuclei. Could there be some small number of nuclear fragments in the mitochondrial fraction? It is possible. Secondly, safranin is hydrophobic: it poorly soluble in water; so it will be embedded in the membrane, and not distributed on both sides in the aqueous phase. Third, a negatively charged succinate (5 mM) will easily pull the positive safranin (1–10 μM) to inside.

In [35] it was stated that safranin suppresses the respiratory chain in the area of NADH-dehydrogenase complex I. This means that it is a poison for the respiratory chain, i.e. its use as a vital fluorescent probe is very problematic.

Merocyanin 540

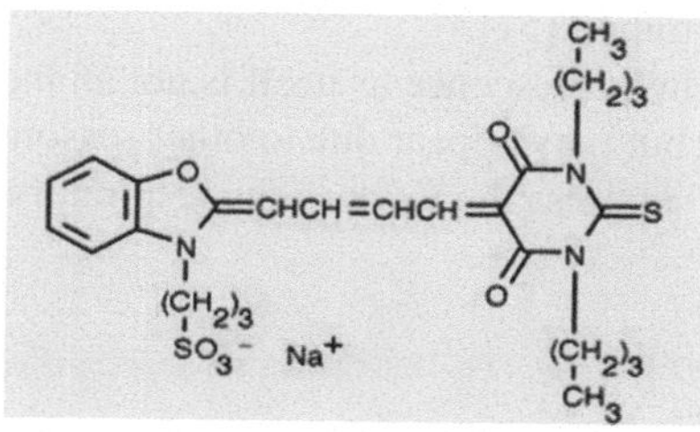

Merocyanin-540 (M540) is usually used as a photo-sensitizer, for example, in photodynamic therapy [50]. In living cells, merocyanin mainly stains membrane structures, interacting mainly with lipid regions. Merocyanin staining in low concentrations does not cause ion transport disturbances and does not prevent the generation of action potential [51].

In aqueous solutions, M540 dissociates into Na^+ cation and anion $M540^-$. Water favors the dimerization of M540. Two maxima are observed in the absorption spectrum of M540: short-wave, belonging to dimers, and long-wave – monomers [52]. M540 is also characterized by the formation of large units (aggregates). For example, the addition of salts to aqueous solutions of this dye leads to the screening of negatively charged groups of molecules M540, increased aggregation and the appearance of a new absorption maximum at a wavelength of 517 nm [53].

Since M540 has a high affinity for lipid phase, but not for proteins, it can be expected that when the dye embedded in the membrane of MC it will not disrupt the respiratory chain. The presence of a large dipole moment (charge) and spectral differences between monomers and dimers, in principle, creates the prerequisites for the use of the M540 as a potential-sensitive fluorescent probe for MCs (see experiments below).

Dis-C3-(5)

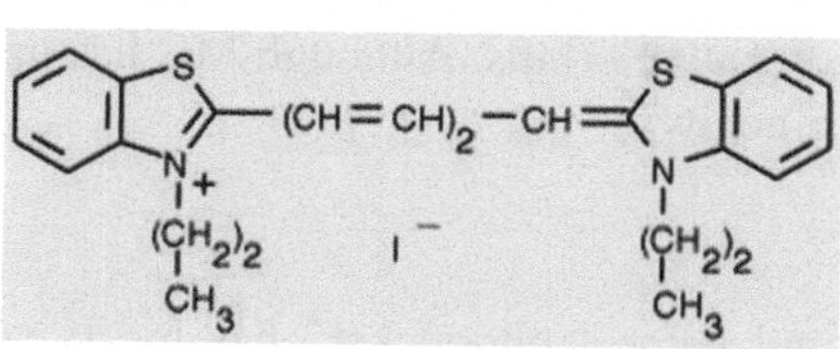

Cyanine cationic dye Dis-C3-(5) is considered as a fluorescent potential-sensitive probe. The maximum excitation is at 620 nm and emission is at 670 nm [37, 54, 55]. This dye was used, in particular, as an indicator of the activity of MCs.

It is believed that Dis-C3-(5) accumulates in the MC when a negative potential appears in matrix as a result of proton transfer during the oxidation of substrates of respiratory chain; in matrix, the dye almost does not fluoresce [55]. Energization of MC by substrates in the suspension is accompanied by some quenching of the dye fluorescence. This was interpreted as the appearance of a negative potential within matrix. Using this technique, the potential on the membrane and its change in damaged MCs was determined [51].

However, the change in fluorescence as itself is not an indication that it is caused by a change in potential, but may appear due to other reasons. By the way, in [35] it is said that cyanine dyes suppress the work of the respiratory chain in the area of the complex I. . .

ANS

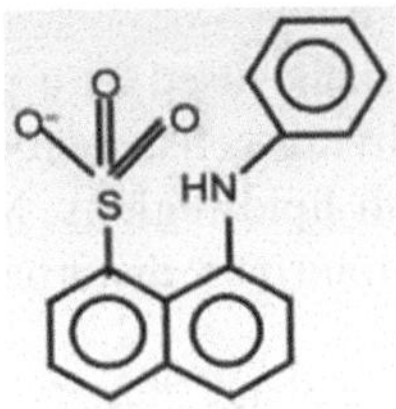

The ANS fluorescent probe (8-anilino-1-naphthalene-sulfonate) is one of the most widely known. An important property of this negatively charged molecule is a sharp increase in the quantum yield of fluorescence during its adsorption into the hydrophobic region of proteins. At the same time, it greatly increases the degree of fluorescence polarization and the lifetime of the excited state. For proteins, the lifetime of ANS is 17 ns [37]. ANS was used, for example, to study the conformation of hemoglobin molecules, to observe local deformations of muscle proteins, to determine the localization of hydrophobic sites in the sarcomere of striated muscle and to track changes in the structure of the membranes of MCs and nerve fibers [51, 54, 55]. It is not at all specific to MCs.

ANS also binds to lipids. Being included in the lipid bilayer membrane, it shows a good quantum yield of fluorescence: 0.29 [51]. At the same time, the life time is small, about 7 ns [37].

Since the ANS probe has a negative charge, its binding to the membranes depends on their surface charge: with the growth of the positive charge, the binding increases, with the growth of the negative – falls. Although MC has negative charge, but we can try to use ANS (see below).

Mitotracker Red

The dye under the brand name "mitotracker" has by its very name declares its specificity for MCs. It is believed that the red mitotracker specifically colors the

MCs with high transmembrane potential [49, 56]. However, any real measurement of values of the potential elsewhere, including [49, 56], is not available. Yes, mitotracker red has an affinity to the intact MCs. But it doesn't just stain only MCs. For example, in [56] the dye stained not only spherical organelles (they are MCs) in cell culture, but also stained actin filaments and other fibrils, which were mistakenly identified as MCs. Mitotracker is not strong better than rhodamine 123 or TMR.

Experimental Verification of Four Probes

When the author [57] tried to use some of the above fluorescent probes to measure transmembrane potential, he got negative results (although at first the data obtained by two undergraduates seemed to be encouraging).

The fraction of MCs isolated by a standard method (at 2 °C in a medium containing 20 mM Tris-HCl and 300 mM sucrose, pH 7.7) from liver of male Wistar rats was added with stirring to solutions of fluorescent probes in 10 mM Tris-phosphate and 150 mM sucrose (pH 7.4). This medium is similar in tonic and other properties to those media that are usually used in measurements of transmembrane potential (see [34]). The concentration of MCs was not higher than 0.3 mg/ml. Probes were taken at a concentration of 0.1–1 μM. The addition of MCs enhanced the fluorescence of the probes due to binding. This amplification was not instantaneous, but developed over time, and in different ways for different probes: from 30 s to 2 min.

When the binding of the probe was over, the addition of succinate was made. The maximum changes in fluorescence response to succinate were observed for M540. Its fluorescence in the suspension of MCs after addition of succinate was increased on 12–15%. Can somebody interpret this observation as a response on energization? Not necessarily. Most likely, the gain of 12–15% is due to additional binding of the probe cation entering to the MC together with the succinate anion.

For all other probes, the fluorescent changes in response to succinate were small and typically did not exceed 5%. How can they be interpreted? Almost as anything, because the fluorescence intensity depends on many factors: the presence in the membranes of aromatic protein residues, iron-sulfur clusters, oxygen, etc. The redistribution of the dye inside the membrane or between the membranes is also important. Local concentrations, from aggregation, light scattering are the others important factors.

The addition of DNP caused in all cases a slight decrease in the fluorescence of all used probes. If DNP drop the potential, the change in fluorescence would have to be much stronger and, moreover – have the opposite direction for cationic and anionic probes. But this is not observed. Apparently, the decrease in fluorescence is caused by the effect of dynamic deactivation of excited probe molecules upon collision with DNP.

The addition of native freshly isolated MCs into medium leads to a slight increase in the fluorescence of the probes: 1.1–1.7 times (Table.4.1). This means that only a small fraction of the probe molecules binds to MC or (and) that the fluorescence quantum yield of the probes varies weakly during binding (which is unlikely; see further experiments with inactivated MCs).

Table 4.1 Fluorescence intensity (ref.units) of potential-dependent probes in diluted suspension of native mitochondria

	Fluor. in medium	+ MCs 0,2 mg/ml	+ succinate 1 mM	+ATP 1 mM	+DNP 50 μm
M-540	100	171	199	207	200
Dis-C3 (5)	100	175	184	176	160
Safranin	100	110	105	103	99
ANS	100	167	168	161	146

Note: Merocyanin: excitation 565 nm/emission 590 nm, Dis-C3-5: 650/680 nm, safranin: 525/585 nm, ANS: 370/450 nm

Table 4.2 Fluorescence intensity (ref.un.) probes in diluted suspension of strongly damaged (inactive) mitochondria

Dye	Fluor. in medium	+ MCs 0,2 mg/ml	+ succinate 1 mM	+ ATP 1 mM	+DNP 50 μM
M-540	100	970	980	990	960
Dis-C3 (5)	100	240	235	230	220
Safranin	100	250	245	240	230
ANS	100	160	150	145	125

The used probes are poorly passed through the outer membrane (despite the huge pores in it) and remain "anchored" in its lipid phase. Therefore, the change in transmembrane potential on the inner membrane will not be able to significantly affect the probe fluorescence. Indeed, the energization by succinate and ATP has very little effect in the case of Dis-C3-5, ANS and safranin (Table.4.1). Only in the case of M-540 the response on succinate is quite pronounced. Whether this is a response to energization has given an experience with died MCs (see below). The addition of DNP led to a marked decrease in fluorescence only in the case of ANS. But this is not the result of a drop in the potential (see below).

To answer the questions asked, control experiments were conducted – with damaged MCs, certainly devoid of any transmembrane potential. The damage was done in two different ways: by double freezing-defrosting or by boiling at 90 ° C for 5 min (died MCs). Similar results were obtained for both methods. Some of them are shown in Table 4.2.

As follows from Table.4.2, the fluorescence of M540 in damaged MCs increased almost tenfold (it is many times more than in the case of intact MCs). This is the result of a sharp increase in the binding of the probes when the membranes are damaged. In the case of Dis-C3-5 and safranin, the intensity in damaged MCs upon dye binding increased by about 2.5 times. This is much more than in the case of intact MCs. So, the damaged MC can bind these probes much better than native ones. And only in the case of ANS there is no significant difference.

It is not any sense at all to talk about transmembrane potential of died MCs. The fluorescence intensity of the used probes gives information not so much about the potential as about the degree of damage of MCs.

In response to succinate and ATP, the fluorescence of the probes in damaged MCs was almost unchanged. Damaged MCs lost their ability to change the fluorescence intensity of probes in response to succinate and ATP. This can be explained by the fact that the free entry-exit of succinate and ATP through the damaged membranes is not accompanied by a symport or antiport of a fluorescent probe. But this has nothing to do with transmembrane potential.

The addition of DNP caused noticeable changes only in the case of ANS. Since the damaged MC has no potential, but the effect of from the addition of DNP exists, therefore, a similar fluorescence decrease in the case of ANS with DNP in native MC is hardly related to the potential. The decrease is due to quenching.

Thus, it can be concluded that such fluorescent probes as merocyanin 540, Dis-C3-5, safranin and ANS do not bind well with native MCs and do not react to the MC energization or de-energization. It is too risky to calculate the transmembrane potential of the internal membrane in response to the addition of succinate, ATP or DNP by the magnitude of the fluorescence change of such probes.

The question of correct measurement of transmembrane potential on the internal mitochondrial membrane remains open.

Conclusion. Widely used fluorescent probes merocyanin-540, Dis-C3-5, safranin and ANS are poorly associated with native MCs and react little to their energization or de-energization. It is not correct to calculate the transmembrane potential of the inner membrane from the fluorescent response at the addition of succinate, ATP or DNP. The results were published in [57].

4.6 Conformational Hypothesis

Although the hypothesis about the possibility of energy transfer from red-ox centers to ATP synthetase through conformational protein transitions was expressed by various researchers, it was most clearly formulated by Boyer [58], who postulated that: (1) the energy released in the respiratory chain is not needed for the act of synthesis ATP molecules from ADP and phosphate, but – to release ATP from the active center, (2) release of ATP from ATPase results from its large-scale conformational transition, (3) the conformational transition in ATPase occurs due to a change to conformations of the respiratory chain enzymes during their red-ox transformations.

The first statement of Boyer was based on the following reasoning. The standard free energy ΔG of ATP synthesis from ADP and phosphate is about 7.5 kcal/mol, so the equilibrium constant is $Kp = 3.3 \times 10^{-6}$. At 5×10^{-3} M phosphate and ADP, approximately 8.3×10^{-11} M ATP will always be present in equilibrium. The development of such a meager amount of ATP does not require an influx of energy.

This amount arises spontaneously. When ADP and phosphate bind to ATPase, their local concentration increases sharply (by several orders of magnitude). Thus, the ratio of ADP to ATP ~ 1 can be obtained on the enzyme. Therefore, the act of phosphorylation does not require additional energy. The following can be added to Boyer's reasoning. Real ΔG strongly depends on pH and concentration of magnesium ions [59]. Enzyme binding of protons and magnesium in the desired region can generally alter ΔG greatly, up to a sign change.

The first postulate of Boyer received a number of vivid experimental evidence. So, Harris et al. [60] showed that isolated bovine heart ATPase contains bound adenine nucleotides: about 3 moles of ATP and 2 moles of ADP per 1 mole of the enzyme. These nucleotides were not separated on Sephadex and other mild treatment methods, but were released upon thermal denaturation of the enzyme. Grubmeyer et al. [61] determined the thermodynamic constants of the stages for the ATPase using radioactively labeled ATP. It was shown that ATP hydrolysis in the active center proceeds without energy release. Consequently, the process of synthesis of ATP from ADP and phosphate does not require energy. Sakamoto and Tonomura [62] observed the synthesis of ATP by soluble ATPase (isolated from MCs) in the presence of dimethyl sulfoxide. The value of *Keqv* reached ~ 1. Synthesis was suppressed by azide. Feldman and Sigman [63] observed the synthesis of ATP by purified ATPase from chloroplasts (in the absence of illumination) with $Keqv = 0.4$. Synthesized ATP remained firmly bound to the enzyme and was not available to hexokinase. It was concluded that the synthesis of ATP does not require energy in the form of proton motive force.

In the mentioned works, it was shown that the reaction of membrane ATP synthesis in MCs and chloroplasts is a simple reversal of ATP hydrolysis, as Boyer postulated. Banks and Vernon [64] were right to a large extent, who earlier stated that Boyer said that ATP synthesis in the catalytic center of ATP synthetase does not require a large influx of energy from the respiratory chain and that the thermodynamic magnitude of the standard free energy ΔG does not have a direct relationship to the coupling.

The results obtained in the experiments were unexpected only for those who initially did not understand that ATPase, like any normal enzyme, should conduct a direct or reverse reaction without attracting external energy, and the direction and speed are determined by concentrations, pH, etc. So the first postulate of Boyer could not be proved.

As for the two other postulates, then with them everything is much more complicated. Although when working with isolated ATPase its conformational changes were registered [65, 66], but they probably do not have a direct relationship to ATP synthesis per se. In model experiments with a single ATPase molecule, rotation of the γ-subunit was observed [65]. It was detected in the range of 1–5 min and correlated with ATPase activity. However, this rotation is too slow. And correlation does not mean cause the effect … In other works (see review [66]), for example, rotation was reported at a speed of 4 turns per second, moreover, rotation took place in discrete jumps of 120 °, each of which was accompanied by the formation of one ATP molecule. However, first, out of hundreds of ATPase molecules, rotation was observed on only a few. Secondly, the rotation as such does not have a direct relationship to the synthesis of ATP. Conformational changes are energy intensive and slow. Enzymes do not work by this way. Enzymes are not much conforming when functioning. Small structural changes usually occur only in the active center [67]. ATP synthetase is hardly an exception. The Boyer model [58], according to which ATPase changes that occur during electron transfer in the respiratory chain, lead to ATP desorption, is true only in that it is important to desorb one ATP molecule to shift the reaction towards ATP synthesis. But for this, no large-scale conformational changes are actually required.

No one has yet been able to trace how the conformational transition in any enzyme of the respiratory chain induces a conformational transition in ATPase with the subsequent release of ATP. Moreover, the issue of conformational transformations in the enzymes of the respiratory chain during red-ox reactions was not confirmed. For example, the X-ray structural data of the oxidized and reduced forms of cytochrome C showed the absence of a significant difference in their conformations [68], contrary to expectations [69]. All changes affected only the active center and did not exceed 0.5 Å [68]. This is true for almost all red-ox enzymes.

Katsumata et al. [70] observed small (about 5%) changes in the protein fluorescence of MCs when adding phosphate, succinate and AMP, which was interpreted as a result of conformational changes. On the other hand, V.I. Sorokovoi et al. [71] found that simultaneously with changes in tryptophan fluorescence of MC proteins, "mirror" changes in NADH fluorescence occur. It was concluded that changes in protein fluorescence are not caused by conformational transitions, but by different efficiency of resonant energy transfer from tryptophan residues to NADH due to NADH $\rightarrow$ NAD reactions. In work [72], on the basis of the fact that changes in the protein fluorescence of MCs with successive addition of succinate, ADP and dinitrophenol in the presence of pyrene and pyrene butyrate were the same as in the absence, it was concluded that the steric availability of tryptophan residues in different functional states membranes does not change significantly, i.e. proteins do not conform strongly.

4.7 Thermal Model of Coupling

With the transfer of electrons through the respiratory chain, about 70% of the energy goes into heat and only 30% is used to generate transmembrane potential and ATP synthesis [73]. It must be emphasized that in warm-blooded animals, the loss of energy into heat is not useless, because it allows them to maintain a constant temperature of the organs.

Heat production in the MC suspension during the oxidation of substrates with oxygen can be detected directly using kinetic microcalorimetry (Fig. 4.5) [74]. Heat production with succinate (entering not only the Krebs cycle, but also directly into the respiratory chain) is much more intense than with hygroxybutyrate.

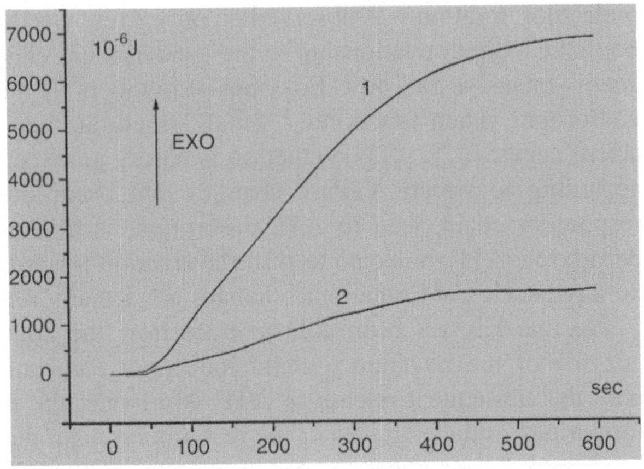

Fig. 4.5 Thermal energy, detected by microcalorimeter during the oxidation of succinate (1) and hybroxybutyrate (2) in mitochondrial suspension at 26 ° C

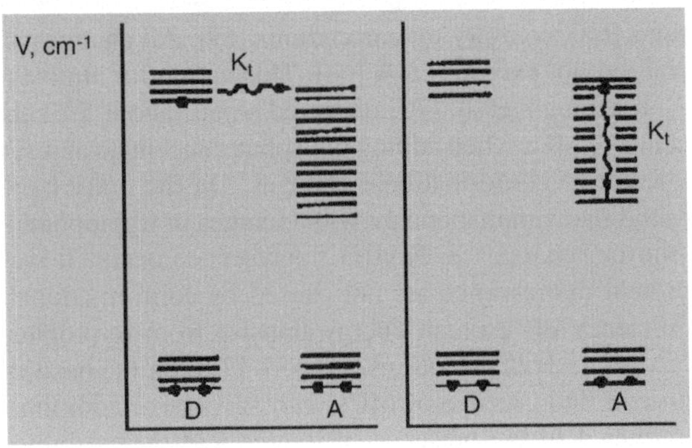

Fig. 4.6 Release of quanta of thermal energy during vibrational relaxation after the reception of an electron by the red-ox center. D is the electron donor, A is the acceptor, Kt is the electron transfer rate constant, Kr is the vibrational relaxation constant

Thermal energy released by intracellular MCs in the form of molecular vibrations and IR radiation is absorbed mainly by cytoplasmic water. Therefore, it is not detected.

Since the transfer of electrons along the respiratory chain is always accompanied by the release of heat, this automatically means that an instantaneous heating of the membrane occurs, i.e. several kcal/mol of thermal energy, sufficient for ATP desorption, appear.

That is why the "thermal" conjugation model was proposed [75]. The essence of the model is that ATP desorption from the active center arises as a result of local heating of the ATPase during the release of "thermal" quanta during rapid vibrational relaxation in the red-ox centers when electrons and protons are received at step by step [75–77]. This model largely coincided with the early idea of K. Straub [78] that the collective oscillations – phonons, arising in the respiratory chain during electron transfer, can cause ATP synthesis (Fig. 4.6).

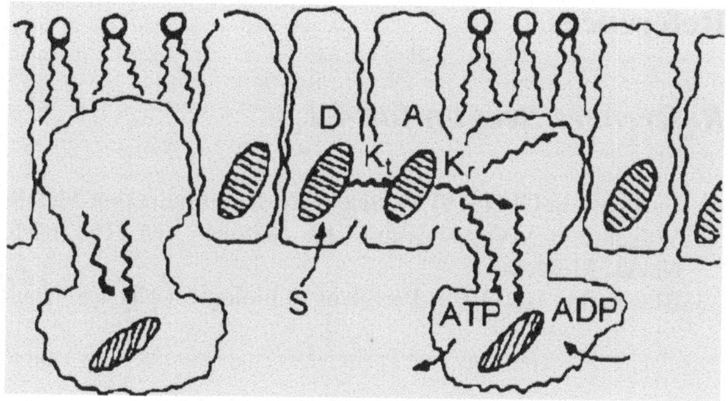

Fig. 4.7 Diagram of a mating membrane operating on the "heat engine" principle. S is the substrate for respiratory chain of MC or quantum of light for chloroplasts

The essence boils down to the following. The difference between the red-ox potentials of the substrates of the respiratory chain and its red-ox centers is 0.3–0.6 eV at each site. The energy quanta released during single electron transfer on the wavenumber scale correspond to about 4000 to 400 cm^{-1}, i.e. are in the middle IR region. Fluctuations of NH-groups at 3300 cm^{-1} (this is 9 kcal/mol) and vibrations of amide-1 at 1600 cm^{-1} can serve as conductors of thermal phonons – collective excitations that, spreading through proteins and reaching ATP synthetase, will lead to the appearance of ATP. In this case, lipid tails are heat insulators, because they are poor in infrared stripes.

Could the membrane proteins to be effective conductors of vibrational excitation? Of course, because peptide bonds are a set of identical or closely spaced oscillators, and the wavelength in the middle IR region is commensurate with the length of the membrane, i.e. one IR-phonon wave can cover several respiratory ensembles at once. Peptide bonds form a set of oscillators that interact strongly with each other: band splitting is observed in the IR spectra of proteins [79]. The strength of an IR oscillator is anisotropic: depending on the orientation, proteins change the intensity of IR bands many times (dichroism) [79]. This means that phonon fluxes in the membranes will have their chosen vectors (Fig. 4.7).

If one could excite the ATP synthetase by external IR light, then ATP synthesis would be observed. Since IR light is strongly absorbed by water, such an experience is impossible. Therefore, Straub proposed [78] to excite with the visible light the electronic absorption bands of cytochromes and other red-ox centers, so that electron excitation, breaking up into many vibrational phonons, reached ATP synthetase and led to ATP synthesis.

Indeed, the irradiation of MC by visible light led to the appearance of ATP [80–82]. Moreover, this experimental result was obtained before the author learned about the prediction of Straub and in general about his phonon model. In the framework of the "phonon-thermal" model, one can explain from a unified standpoint the ATP synthesis in MC and ATP photosynthesis in chloroplasts.

References

References Section for 4.1

1. Lichtenstein GI (1979) Multicore oxidation-reduction Metalloenzymes. Nauka, Moscow
2. Volkenstein MV, Golovanov IB, Sobolev VM (1982) Molecular orbitals in enzymology. Nauka, Moscow
3. Blumenfeld LA (1974) Problems of biological physics. Nauka, Moscow

References Section for 4.2

4. Blumenfeld LA (1974) Problems of biological physics. Nauka, Moscow
5. Wolkenstein MV, Golovanov IB, Sobolev VM (1982) Molecular orbitals in enzymology. Nauka, Moscow
6. Orchin M, Jaffe G (1969) Loosening orbitals. Mir, Moscow
7. Dmitriev IS, Semenov SG (1980) Quantum chemistry. Atomizdat, Moscow

References Section for 4.3

8. Cox M, Nelson D (2017) Princ Lehninger's Biochem 2
9. Malenkova IV, Kuprin SP, Davydov RM, Blumenfeld LA (1980) Dokl Acad Sci USSR 252 (3):743
10. Vekshin NL (1988) Coupling of ATP synthetase with membrane electron and energy carriers; thermal model. Preprint of NCBI, Pushchino

References Section for 4.4

11. Mitchell P (1961) Nature 191:191.144
12. Mitchell P (1966) Biol Rev 41:41.445
13. Skulachev VP (1972) Transformation of energy in biomembranes. Nauka, Moscow
14. Skulachev VP (1989) Energetics of biological membranes. Nauka, Moscow
15. Nichols DD (1985) Bioenergy. Introduction to chemiosmotic theory. Mir, Moscow
16. Mitchell P, Moyle J (1969) Eur J Biochem 7:471
17. Blumenfeld LA (1974) Problems of biological physics. Moscow, Nauka
18. Vekshin NL (1988) Coupling of ATP synthetase with membrane electron and energy carriers; thermal model. Preprint of NCBI, Pushchino
19. Maloff BL et al (1978) J Cell Biol 78(1):214
20. Komai H et al (1976) BBRC 69:697
21. Knowles AF, Racker E (1975) J Biol Chem 250:3.1949
22. Vale M et al (1976) Biochem J 156:3.239
23. Yaguzhinsky LS et al (1979) Biophysics 24(6):1100
24. Reid RA, Moyle J, Mitchell P (1966) Nature 212(5059):257
25. Jagendorf AT, Uribe E (1966) PNAS 55:170

26. Malenkova IV, Kuprin SP, Davydov RM, Blumenfeld LA (1980) Dokl Acad Sci USSR 252 (3):743
27. Blumenfeld LA, Goldfeld MG, Dmitrovsky LG (1978) Biophysics 23(3):549
28. Racker E, Stoeckenius WJ (1974) Biol Chem 249:662–663
29. Hartmann R, Oesterhelt D (1977) Eur J Biochem 77(2):325
30. Michel H, Oesterhelt D (1980) Biochemistry 19(20):4.4615
31. Lehninger A (1974) Biochemistry. Mir, Moscow
32. Eremeev SA, Yaguzhinsky LS (2015) Biochemistry 80(5):682–688

References Section for 4.5

33. Skulachev VP (1989) Energetics of biological membranes. Nauka, Moscow
34. Bodrova ME, Dedougou VI, Mokhov EN (2000) Biochemistry 65(4):562
35. Ghelli A, Benelli B, Esposti MD (1997) J Biochem 121:746
36. Freedman JC, Novak TS (1989) In: Fleisher S (ed) Methods in Enzymol, vol 172. Academic, New York, p 102
37. Vekshin NL (2008) Fluorescence spectroscopy of biopolymers. Photon, Pushchino
38. Lakowicz J (1999) Prinsiples of fluorescence spectroscopy. Plenum, New York
39. Johnson LV, Walsh ML, Chen LB (1980) PNAS 77(2):990
40. Jahnson VL, L M et al (1981) J Cell Biol 88(3):526
41. James TW, Bohman R (1981) J Cell Biol 89(2):256
42. Ozernyuk ND (1978) Growth and reproduction of mitochondria. Nauka, Moscow
43. Luzikov VN (1980) Regulation of mitochondrial formation. Nauka, Moscow
44. Belyakovich AG (1990) Study of mitochondria and bacteria using the salt of tetrazolium p-NTV. ONTI, Pushchino
45. Gear ARL (1974) J Cell Biol 249:3628
46. Kulik AV, Nekrasova OE, Minin AA (2006) Biol Membr 23(1):42
47. Perevoshchikova IV, Zorov DB, Antonenko YN (2008) Biochim Biophys Acta Biomembr 1778(10):2182
48. Perevoshchikova IV, Sorochkina AI, Zorov DB, Antonenko Y (2009) Biochemistry 74(6):663
49. Perevoshchikova IV (2009) PhD-dissertation in biological sciences. Moscow State University, Moscow
50. Kozhinova EA (2007) Disssertation of PhD biology sciences. RSMU, Moscow
51. Vladimirov YA, Proskurnin EV (2007) Lectures on medical biophysics. Moscow state University
52. Sikurova L, Cunderlikova B (1997) Spectrochem Acta Part A 53(293)
53. Sikurova L et al (1995) Anal Chem Acta 303:79
54. Haugland R (2001) Molecular tests. Handbook (USA)
55. Vladimirov YA, Dobretsov GE (1980) Fluorescent probes in the study of biological membranes. Mjscow, Nauka
56. Chernyak BV (2010) Dissertations doctor biology sciences. Moscow state University
57. Vekshin NL (2013) Biophysics 58(6):1074–1080

References Section for 4.6

58. Boyer P (1974) Dynamics of energy transmit membranes (Ernster L (ed)). Elsevier, Amsterdam, p 289

59. Matthies M, Zundel G (1979) J Inorg Biochem 10(2):109
60. Harris D et al (1973) BBA 314:149
61. Grubmeyer C et al (1982) J Biol Chem 257:12092
62. Sakamoto J, Tonomura Y (1983) J Biochem 93:1601
63. Feldman R, Sigman DJ (1982) Biol Chem 257:1667
64. Banks B, Vernon C (1970) J Theor Biol 29:301
65. Bald D, Noji H, Yoshida M et al (2001) J Biol Chem 276(43):39505–39507
66. Romanovsky YM, Tikhonov AN (2010) Usp Phy Sci 180(9):931–956
67. Volkenshtein MV, Golovanov IB, Sobolev VM (1982) Molecular orbitals in enzymology. Nauka, Moscow
68. Takano T, Dickerson R (1981) J Mol Biol 153:95
69. Blumenfeld LA (1974) Problems of biological physics. Moscow, Nauka
70. Katsumata Y, Miyazi A, Ozawa TJ (1970) Biochemist 68:423–435
71. Sorokova VI, Klebanov GI, Vladimirov YA (1972) Mol Biol 6:705–711
72. Vekshin NL (1983) The use of nonradiative energy transfer to study biomembranes. Preprint. Pushchino

References Section for 4.7

73. Lehninger A (1974) Biochemistry. Mir, Moscow
74. Kotelnikov GV, Moiseeva SP (2016) Sci Pribor 26(3):3–9
75. Vekshin NL (1988) Conjugation of ATP synthetase with membrane carriers of electrons and energy; thermal model. Pushchino, NCBI
76. Vekshin NL (1990) Comm Mol Cell Biophys 7(1):17–25
77. Vekshin NL (1991) Biophysics 36(6):994–999
78. Straub KD (1974) J Theor Biol 44:191
79. Chirgadze YN (1972) Sat results of science and technology, Molecular biology-1. Moscow, VINITI, pp 9–60
80. Vekshin NL, Mironov GP (1982) Abstracts of the All-Union Biophysical Congress. V1. Moscow
81. Vekshin NL (1983) Physico-chemical bases of cell functioning. ONTI, Pushchino, pp 82–86
82. Vekshin NL (1991) Mol Biol 25:54–59

Chapter 5
Photoreactions

5.1 Photo-Respiration and Photo-Induced ATP Synthesis

Lighting of MCs can lead to oxygen consumption — photorespiration [1], as well as photo-induced ATP synthesis [2, 3]. The scheme of the pilot plant is shown in Fig. 5.1.

When lighting suspension of MCs (from rat liver) is carried out, the oxygen consumption arises (even in the absence of added substrates, Fig. 5.2) [1, 4]. It stops immediately after turning off the light. The photorespiration inhibited by sodium azide, which acts not only as a blocker of the respiratory chain at the level of cytochrome oxidase, but also as a deactivator of photoexcited flavins in flavoproteins.

Under illumination of MCs, there are processes involving flavoproteins:

$$Fla + h\nu \rightarrow Fla* \rightarrow Fla\hat{}$$

$$Fla\hat{} + O_2 \rightarrow Fla + O_2*$$

$$O_2 * + Lip \rightarrow LipO_2$$

$$Sub + Fla\hat{} \rightarrow FlaH_2 \rightarrow Fla + H_2O_2$$

Here hν is quantum of blue light, Fla is flavoprotein, Fla$*$ is its excited singlet state, Fla ^ is a triplet state, Fla-H_2 is a reduced form of flavin, Lip is lipids, $LipO_2$ is their oxidation products, Sub is a substrate – ketoglutarate or NADH [1, 4]. Oxygen consumption under the light irradiation occurs (a) due to lipid peroxidation, (b) due to activation of the oxidation of substrates by the respiratory chain – photorespiration.

The appearance of photorespiration in isolated MCs was confirmed by Konev and Rudenok [5], moreover, the photo-consumption of oxygen was observed even under the conditions of cyanide block of cytochrome oxidase. Hence, it is unlikely that photorespiration to be caused by the photo-activation of cytochrome oxidase.

N. Vekshin, *Biophysics of Mitochondria*,
https://doi.org/10.1007/978-3-030-33853-4_5

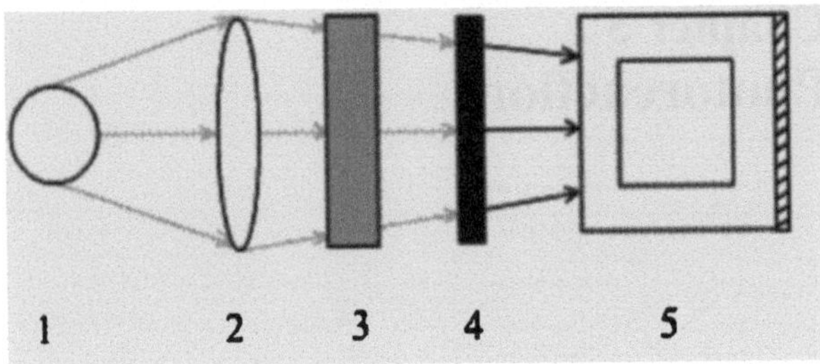

Fig. 5.1 Installation for photo-irradiation. 1 is a lamp, 2 is a lens, 3 is a heat filter, 4 is a light filter, 5 is a holder for a cuvette (with a window and a mirror)

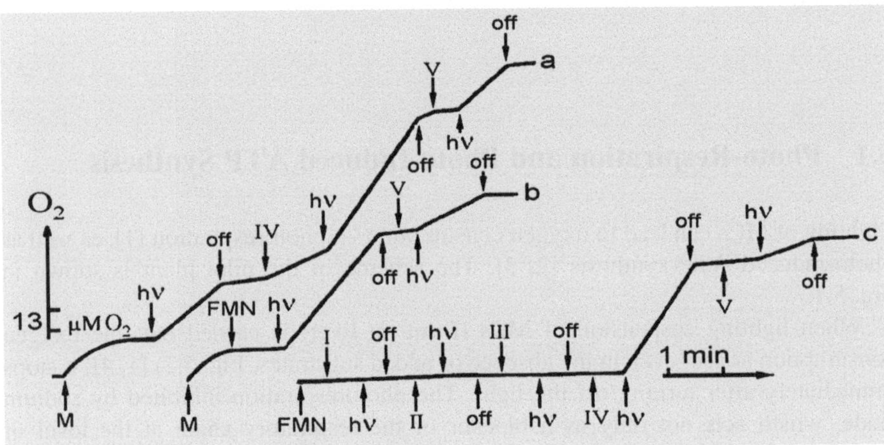

Fig. 5.2 Photo-induced oxygen consumption, measured polarographically in a suspension of hepatic mitochondria (under conditions of thermo-stating and mixing in a sealed transparent cell): (**a**) – under the light of the SVD-120A lamp (plus heat filter), (**b**) – the same in the presence of added FMN irradiated with the 436 nm line of the lamp, (**c**) – photochemical reaction upon irradiation of an FMN solution (without mitochondria) with the 436 nm line with the addition of substrates (5 mM each), I – succinate, II – glutamate, III – oxybutyrate, IV – ketoglutarate; Y – azide

In [6], under the action of light, the restoration of cytochromes of mitochondrial complexes incubated with flavins and EDTA added was observed.

The addition of FMN led to a sharp acceleration of oxygen consumption under the light (Fig. 5.2). In this case, as in [6], direct photochemical reactions of FMN with oxygen in solution made a significant contribution to the total photo-induced consumption of oxygen.

The introduction of succinate, glutamate or hydroxybutyrate did not affect the oxygen consumption, but the addition of NADH or ketoglutarate led to abrupt activation, which was caused mainly by direct photochemical reactions in solution, rather than by photorespiration of MCs [1, 4].

In experiments [1, 4], rotenone and antimycin only slightly reduced the photo-respiration of native MCs and did not affect the FMN-induced photo-consumption of oxygen at all.

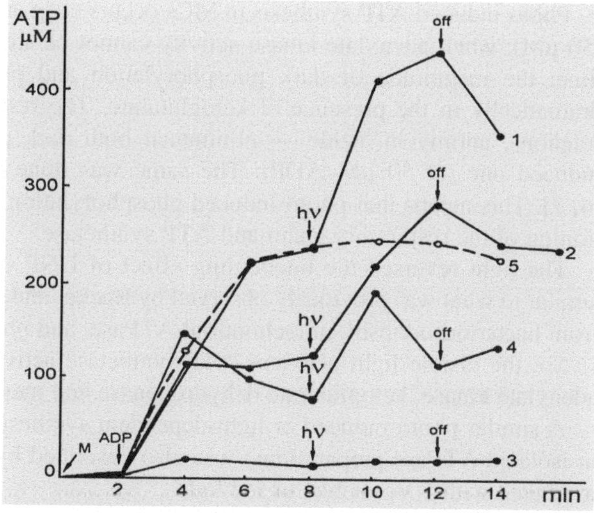

Fig. 5.3 Changes in ATP concentration in mitochondrial suspensions in the dark and upon blue light. 1 – added 1000 μm ADP; 2 – 700 μM ADP added; 3 – 200 μM ADP added; 4 – in the presence of 3 mM ketoglutarate, 200 μM ADP was added; 5 – 1000 μM ADP added, no light. Samples were taken every 2 min and conservated by TCA, after which the amount of ATP was measured (by the luciferase method)

In addition to photoreactions when light is absorbed by flavoproteins (at 450 nm) and cytochromes (the intense Soret band at 420 nm) the MCs can spend the electron excitation energy into set of thermal vibrational quanta. At the same time, almost every mole of absorbed photons in the membrane gives in the balance of ΔH ~ 60 kcal of thermal energy.

$$Fla + h\nu \rightarrow Fla* \rightarrow Fla + \Delta H_1(kcal/mol)$$

$$Cyt + h\nu \rightarrow Cyt* \rightarrow Cyt + \Delta H_2(kcal/mol)$$

This should lead to a strong local "warming up" of the membrane and acceleration of ATP synthesis. Indeed, it was observed [2–4, 7].

When large quantities of ADP are added to the MC suspension in the absence of added substrates, the usual dark phosphorylation of half of the added ADP to ATP occurs, which is mainly due to adenylate kinase activity and ATP synthetized activity. The inclusion of blue light induces a sharp rise in the level of ATP (Fig. 5.3) [3, 4, 7]. After 2–4 min of illumination, the amount of ATP ceases to increase or even decreases. Turning off the light leads to a decrease in the level of ATP. It is caused by an increase in ATPase activity, apparently due to the oxidation of membrane lipids by singlet oxygen. The introduction of succinate accelerated dark phosphorylation of ADP to ATP, but did not affect photo-induced phosphorylation. The introduction of ketoglutarate against 200 μM ADP led to a sharp increase in both dark phosphorylation and photo-induced one [4, 7]. The maximum of the spectrum of action of light is about 436 nm [4]. The spectrum of action is close to the absorption bands of flavins and cytochromes.

Photo-induced ATP synthesis in MCs occurs even at low concentrations of ADP (50 μM), when adenylate kinase activity cannot be significant [4, 7]. At the same time, the magnitude of dark phosphorylation and photo-induced one increases dramatically in the presence of ketoglutarate. The respiratory chain inhibitors — rotenone, antimycin, azide — eliminated both dark phosphorylation and photo-induced one (at 50 μM ADP). The same was done by oligomycin and DCCD [4, 7]. This means that photo-induced phosphorylation is associated with the functioning of the respiratory chain and ATP synthetase.

The light reversed the uncoupling effect of DNP on MCs [4, 7], in a manner similar to what was previously observed by Recker and Stokenius [8] on "chimeras" from bacteriorhodopsin, mitochondrial ATPase, and phospholipids.

So, the visible light activates ATP-synthetase activity of MCs on three levels: adenylate kinase, ketoglutarate dehydrogenase and membrane ATP-synthetase.

A similar photo-induced or light-dependent synthesis of ATP in MCs (and even in isolated ATPase preparations) was also described by other authors [9–11] upon irradiated with UV, visible, or red light.

Photorespiration and ATP synthesis in the illuminated MCs were found in cells of plant seedlings [12–14]. In MCs of plants, these processes, apparently, perform a certain physiological role [14].

Photodynamic aspects of the action of laser irradiation on various non-photosynthetic cells were considered in [15].

In [16], it was shown that near infrared light (600–900 nm) is able to modulate cytochrome oxidase activity and ATP synthesis in brain cells.

5.2 Photo-Desorption from Surface

Many biochemical studies require the use of a minimal amount of a sample – cells or their organelles. And at the same time it is often necessary that the sample does not spoil for a long time, retains enzymatic activity.

One of the ways to solve these problems is to use a smear of organelles or cells on a plate. The procedure for obtaining an immobilized sample from organelles (or cells) in the form of a thin dried smear usually does not lead to a loss of the reductase activity of membrane dehydrogenases [17]. This activity is measured photometrically or fluorimetrically using artificial electron acceptors.

In a photometer, and especially a fluorometer, the passage of intense light through a sample can, in principle, lead to a variety of photo-effects. For example, during UV irradiation of proteins, photo-conformational relaxation occurs [18], up to denaturation and aggregation [19]. Under UV irradiation of dehydrogenases, the NADH molecule is desorbed from the active site due to the instantaneous "warming up" of the protein globule [18]. In the MC suspension, powerful blue light activates respiration and ATP synthesis [18]. During the photo-degradation of flavin in NADH dehydrogenase, the suppression of oxidase activity occurs, but reductase activity towards artificial acceptors remains [20].

In work [21], a highly sensitive fluorimetric method for detecting dehydrogenases in a thin smear of membranes or cells on a transparent plate was proposed. It was found that the passage of too intense light through a smear of membranes or cells can lead to desorption of them into the solution. Below, a description of this photo-desorption effect is given on the example of hepatic MCs.

MCs were isolated from rat liver, based on [22–24]. A smear preparation was performed according to [17] with some modifications. A micro-droplet (2 μl) of a thick suspension of MCs was placed on a degreased quartz plate and smeared by a cover glass to very thin even layer. Then the smear was dried for 2 min. In the control experiments, the dried smear was immersed for 1 min in isopropanol, ethanol, or acetone for stronger fixation; then the sample was washed with water and dried again.

Desorption of MCs from the plate into a solution was detected on a Perkin-Elmer spectrofluorimeter as reduce of the tryptophan fluorescence of a smear after irradiation with intense light on the same instrument. Irradiation was carried out for 10–15 minutes from a 150-W xenon lamp through a wide monochromator slit of 20 nm.

Protein tryptophan fluorescence was detected at 340 nm (slit 4 nm) at 280-nm excitation through a narrow slit (2 nm). During irradiation and measurement, a quartz 1-cm cuvette was rigidly, without backlash, fixed in the instrument holder, and a quartz plate with a smear was placed diagonally in the cuvette as described in [21] (Fig. 5.4), and was also well fixed. The cuvette was filled with 3 ml of a solution of 250 mM sucrose and 20 mM Tris-HCl (pH 7.4). NADH: pNTP reductase reaction with formazan accumulation in MCs was performed as described in [21].

Although the UV absorption band of MCs in a smear is strongly hypochromised compared to MCs in suspension of the same concentration (the causes of hypochromism are discussed in [18]), however, MC fluorescence from smear is many times higher than from suspension. The reason for this is that the excitation light is focused by the lens into the center of the cuvette and is collected by the second lens also from the center through which the plate passes and where the smear is concentrated.

With the correct diagonal position of the smear plate, the maximum fluorescence intensity is achieved; moreover, the excitation light does not enter the recording channel (reflected in the opposite direction) and does not interfere with the measurements.

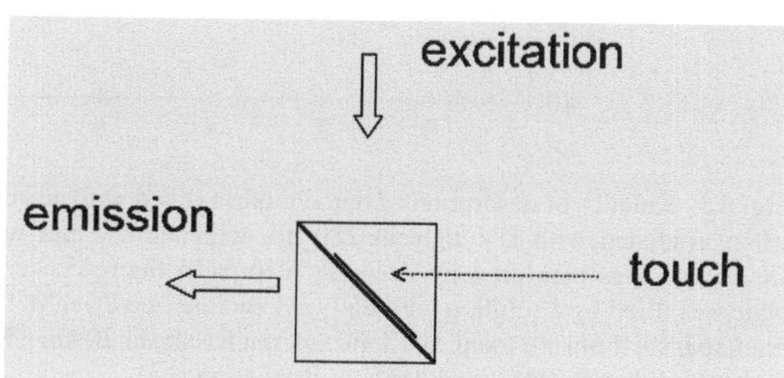

Fig. 5.4 Cuvette with a diagonal plate (top view). Smear of MCs is on the back side of the plate

The presence of a large difference in the fluorescence intensities of MCs on the plate and in the solution determines the possibility of successful detection of the MC sorption-desorption processes.

Figure 5.5 shows the effect of UV irradiation on the intensity of tryptophan fluorescence of the MC smear under different conditions. The observed decrease of smear fluorescence over time is due to gradual photo-desorption of MCs, which is accompanied by the appearance of protein fluorescence in the solution (this is detected after removing the plate).

Slight photo-desorption of MCs arises also upon irradiation by visible light (Table 5.1). The photo-desorption rate strongly depends on the spectral composition of irradiation. It was maximal at tryptophan absorption band – at 280 nm. Note that this occurs despite the fact that the xenon lamp of the device gives much less light in the UV region than in the visible region.

As can be seen from Fig. 5.5, the greatest decrease in tryptophan fluorescence of the MC smear proteins due to photo-desorption is achieved with wide slits of the exciting monochromator – 20 nm. At 2 nm slits, photo-desorption is much smaller (spontaneous "dark" desorption makes a significant contribution), i.e. the photo-desorption rate depends strongly on the intensity of irradiation. In control experiments (smear incubation in the dark), the decrease in fluorescence was much slower than during irradiation, i.e. spontaneous desorption of MCs from the plate into the solution occurs slowly (Fig. 5.5).

All of the above applies only to freshly isolated MCs and thin smears. A smear of a long-stored MC suspension was poorly absorbed onto the plate (even as a very thin

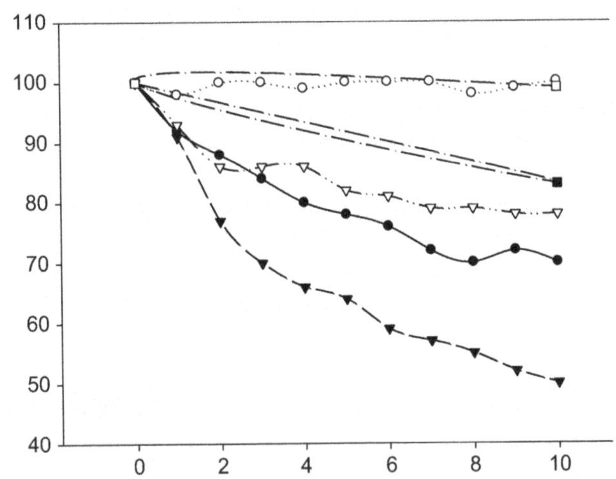

Fig. 5.5 Kinetics of desorption of mitochondria (fixed with acetone and non-fixed) in the dark and when irradiated with UV light at 280 nm with narrow and wide slits of the monochromator. Desorption was measured by decrease of protein fluorescence. Cuvette with smear touch on a plate was filled by a solution with 250 mM sucrose and 20 mM Tris-HCl (pH 7.4). The touch was: not fixed, slit 2 nm (•), fixed, slit 2 nm (o), not fixed, slit 20 nm (▼), fixed, slit 20 nm (∇), not fixed, without irradiation (■), fixed, without irradiation (□)

Table 5.1 Photo-desorption of mitochondria (without special fixation of the MC smear) depending on the illumination wavelength

Irradiation wavelength	Photo-desorption (%)
280 nm	45
310 nm	32
360 nm	24
410 nm	5
510 nm	5

Note: Protein fluorescence (excitation slit 2 nm) was quickly measured after 10 min of irradiation (excitation slit 20 nm). Photo-desorption is given minus spontaneous desorption

Table 5.2 Spontaneous desorption of mitochondria from a smear into a solution, depending on conditions

Mitochondria:	Desorption (%) during 10 min
Unfixed, in a thin smear	10
Fixed by isopropanol	1
Fixed by ethanol	10
Fixed by acetone	0
Long keep	80
Thick smear	95

layer) and was easily spontaneously desorbed. It means that the adhesion properties of the outer membrane are violated during long-term storage. In thin smears, each MC is attached to the plate. Thick smears (where most of MCs do not come into contact with the plate, but with each other) "sit" on the plate poorly and easily go spontaneously into the solution without any irradiation.

When the smear was pre-fixed with isopropanol, spontaneous desorption did not occur at all, and noticeable photo-desorption was observed only with large monochromator slits (Fig. 5.5). Fixing the smear with ethanol or acetone was less reliable than isopropanol (Table 5.2). Ethanol did not provide reliable fixation of the smear: when moving from one medium to another, the smear "crawled" stronger than in the case of acetone or isopropanol. Acetone has a marked absorption at 280 nm. Due to the preservation of traces of acetone in a smear, a long drying was required.

Photo-desorption takes place due to local heating of MC membranes of smear at the interface with the plate upon absorption of light. This is not heating of the solution in which the smear plate is located. During the experiment, the solution in the cuvette was heated after 10 min by no more than 0.5 degrees. By other words, only a prompt local "warming up" of membrane proteins took place. The quantum yield of tryptophan fluorescence of mitochondrial proteins is low [18], it does not exceed 20%, i.e. at least 80% of the absorbed energy is spent without emission. This energy is spent in heat. A photon of 280 nm gives thermal energy of about 100 kcal/mol. This energy is enough to break several dozen hydrogen and adhesive bonds, which can lead to detachment of MC from quartz plate.

If the mitochondrial smear will paint by formazan formed during pNTP reduction during the NADH: pNTP reductase reaction, and then irradiate in the absorption band of formazan at 540 nm, the MC desorption is very low (Table 5.3).

Table 5.3 Photo-desorption of mitochondria from a smear during irradiation in the presence of formazan

Irradiation wavelength	Photo-desorption of mitochondria (%)
280 nm	45
540 nm	10

This means that heating of each MC inside (through formazan deposited in matrix [17]) during irradiation does not play a significant role. From this experience, it can also be concluded that the passage of intense light in photometers or fluorimeters through a smear of organelles or cells will not prevent the use of tetrazolium dyes for measuring of dehydrogenase activities.

5.2.1 *Conclusion*

Photo-desorption of MCs adsorbed on a quartz plate was discovered and studied. The speed of photo-desorption of MCs from the plate into a solution depends on the length of the light wave, the intensity of irradiation and its duration. Maximum speed is achieved by UV irradiation in the tryptophan absorption band of proteins. UV photo-desorption is caused by a local photo-thermal effect: heating of photo-excited surface membrane proteins that attach MC to the plate. Preliminary fixation of a smear with isopropanol, ethanol or acetone dramatically reduces photo-desorption. The data obtained are important for the development of the technology of immobilization of MCs or cells for measurements of enzymatic activities and for biosensors. The results were published in [25].

5.3 Flavin and Ubiquinone Are Not Involved in the Transfer of Electrons to Artificial Acceptors

The NADH dehydrogenase complex I can catalyze the reduction of ubiquinone and proton transfer across the membrane coupled with ATP synthesis [26]. The NADH dehydrogenase complex of the bull heart MCs consists of 42 subunits and contains flavin (FMN or FAD) and six iron-sulfur clusters, one of which (Fe_4S_4) is nearby [27–29].

NADH can donate electrons through the enzyme complex not only to ubiquinone, but also to other acceptors. The catalytic properties of NADH dehydrogenase in MCs and also in sub-mitochondrial particles or isolated detergent preparations are studied using artificial acceptors [30–33], for example, ferricyanide, pNTP, DCPIP, etc. However, only few people wondered whether the paths of electron transfer from the substrate to artificial acceptors by the same pathway like to ubiquinone or cytochromes.

Such acceptors as ferricyanide cause the oxidation of NADH by the enzyme bypassing ubiquinone [34]. There are indications that the reduction of ferricyanide occurs on high-potential iron-sulfur clusters [35], in a completely different area than the restoration of ubiquinone [34]. Ubichinol is considered to be a reducing agent of such acceptors as pNTP and DCPIP [35], but it is not true.

It is believed that flavin and iron-sulfur clusters are involved in the transfer of electrons from NADH to ferricyanide, and all red-ox components of the relay: flavin, iron-sulfur clusters, ubiquinone [26, 27, 32, 33] are involved in the transfer of electrons to pNTP or DCPIP. The basis for such schemes was the difference between the red-ox potentials of these red-ox substances. However, the difference in itself is not evidence of the realization of the path taken by the electron. Moreover, in [30], it was found that the rate of ferricyanide-dependent oxidation of NADH in one of the enzyme subunit fractions, depleted of flavin (but rich with iron), is many times higher than in the flavin-rich enzyme.

It was interesting to study the role of flavin and ubiquinone in the enzymatic oxidation of NADH, activated by artificial acceptors – ferricyanide, pNTP and DCPIP. The work was conveniently carried out with PMCs <0.2 microns – the smallest mitochondrial organelles that do not possess such strong light scattering as MCs.

The PMC fraction containing membrane NADH dehydrogenase was obtained from the suspension of MCs of the bovine heart by filtration through 0.2 μm Millipore filters (Fig. 5.6). Another advantage of using PMCs <0.2 microns is that these organelles, in contrast to the suspension of heavy MCs, are not so quickly deposited to bottom.

To measure the activity of NADH-dehydrogenase, the PMC fraction in a hypotonic medium — 5 mM or 20 mM Tris-phosphate buffer (pH 7.5 at 20 ° C) — was placed in 0.5- or 1-cm quartz cuvettes at "Specord UV Vis" or "M-40". NADH oxidation was determined by a change in optical density at 340 or 360 nm, and also acceptor recovery was measured: at 410 nm – ferricyanide, at 600 nm – DCPIP and at 530 nm – formazan, which is formed during pNTP restoration. The protein concentration was determined by the UV express method [36].

The flavin fluorescence was recorded at 525 nm (excitation – 450 nm) on the "SLM-4800". PMCs were placed in a mirror cuvette, allowing to enhancing the signal [37].

To carry out photochemical experiments, the cuvette was placed in a metal holder with a front window and a rear mirror. The cuvette was thermostatically controlled at 14 ° C. Illumination of flavins in PMCs (in the control – FMN solution) with a line of 436 nm of lamp SVD-120A through filter SZS-8 and a heat filter was carried out for 1–60 min, taking samples during irradiation for subsequent measurements of a) enzymatic activity and b) destruction of flavins.

Intensive long-term blue irradiation of free FMN in a solution causes its irreversible destruction; the absorption band at 450 nm disappears (Fig. 5.7) [37]. Also some changes at the absorption in the UV region take place. The addition of iron ions, forming a complex with FMN, slows down the photodestruction (Fig. 5.7). This indicates a high affinity of FMN for iron and a strong electron interaction between them.

Fig. 5.6 Obtaining the proto-mitochondrial fraction from a diluted suspension of mitochondrial organelles using calibrated Millipore filters

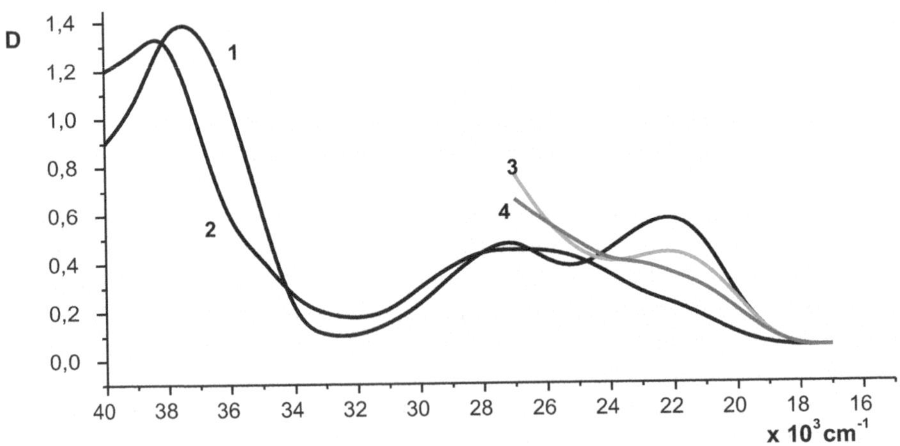

Fig. 5.7 Absorption spectrum of FMN in water: before irradiation (1), after (2), in the presence of 280 μM FeCl3 before irradiation (3), the same after irradiation (4). The exposure time is 12 min

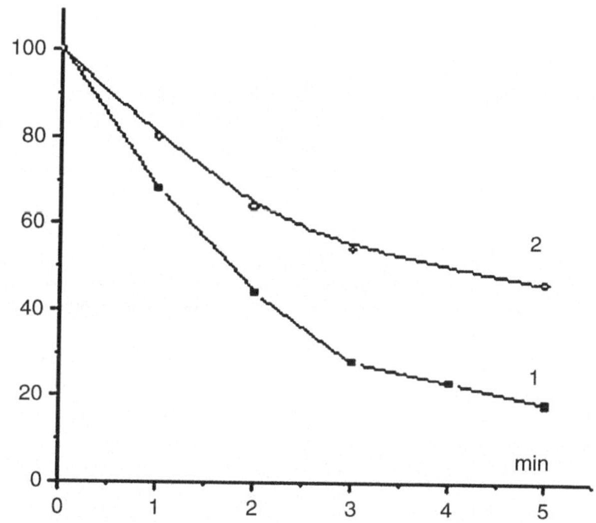

Fig. 5.8 Destruction of free FMN (1) and flavins in proto-mitochondria (2) during 5-min irradiation. It was measured (after finish of irradiation) by reducing in the flavin fluorescence excited at the band 450 nm and detected at 525 nm

Photo-destruction of FMN is accompanied by disappearance of its fluorescence. This was used to detect the destruction of flavin in NADH dehydrogenase in PMCs (Fig. 5.8).

Short-term (1–2 min) irradiation of MCs with blue light may be accompanied by flavin-dependent oxygen consumption and ATP synthesis [37, 38]. One of the claimants to participate in this is NADH dehydrogenase.

Flavin in NADH dehydrogenase makes an overwhelming contribution to the total flavin fluorescence of PMCs from the bovine heart [28]. Destruction of flavin in NADH dehydrogenase of PMCs during irradiation occurs significantly slower than free FMN in solution (Fig. 5.8). In the enzyme next to the flavin, there is some group that effectively deactivates its excited state. Apparently, this group is the nearest cluster Fe_4S_4 [28]. The electrons of the d-orbital of iron extend longer 10 Å [39], i.e. the interaction between iron and flavin can occur at such a distance between the centers.

Some of the PMC flavins are not subject to photo-destruction: after 5–10 min of illumination, the curve reaches a plateau (Fig. 5.8). This is due to the fact that the flavin fluorescence of PMCs is caused not only by NADH dehydrogenase, but also by several FAD proteins. FAD is more photo-resistant than FMN.

The destruction of the proto-mitochondrial flavin, observed under illumination, was not accompanied by changes in tryptophan fluorescence (not shown in a figure), which indicates the absence of conformational changes in proteins. However, destruction led to a decrease in the oxidation rate of NADH in the respiratory chain, measured after irradiation (Fig. 5.9). This seems to confirm that flavin is essential for NADH: ubiquinone reductase activity. At the same time, the difference in the amount of photo-destruction and the amount of activity attracts attention. For

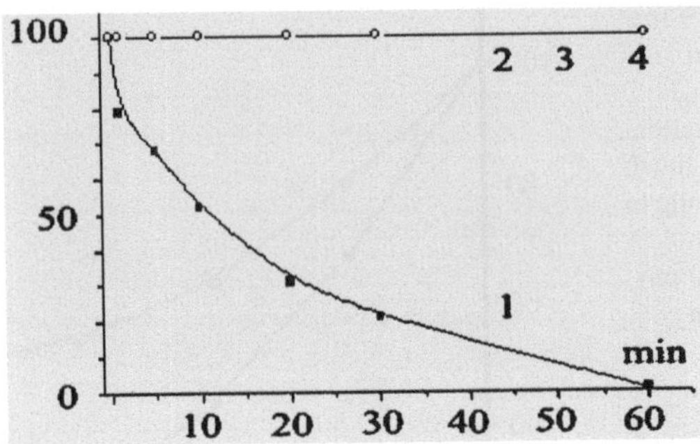

Fig. 5.9 Dependence of the activity of NADH dehydrogenase in PMCs on irradiation time: (1) NADH: ubiquinone-reductase activity, (2) NADH: ferricyanide-reductase, (3) NADH: pNTP-reductase, (4) NADH: DCPIP-reductase. Activities were measured by changes in optical density of NADH or acceptors (in %)

example, after 5-min irradiation, the flavin fluorescence drops by 50%, and NADH: ubiquinone-reductase activity decreases by only 30% (Figs. 5.8 and 5.9). There are two reasons for this: a) in PMC there is a pool of flavins not belonging to NADH-dehydrogenase, b) flavin is not strictly necessary for enzymatic oxidation of NADH.

The photo-destruction of flavins did not affect the ferricyanide-activated NADH oxidation at all (Fig. 5.9). This means that flavin is certainly not a participant in the transfer of electrons from NADH to ferricyanide. Similar results were obtained with pNTP and DCPIP instead of ferricyanide (Fig. 5.9).

When these acceptors were added, the flavin fluorescence of PMCs did not change (it was measured without NADH and irradiation). This means that acceptors bind far from flavin. It means that they can hardly accept electrons from the reduced form of flavin.

Dithiotretitol (DTT), which strongly inhibits ferricyanide-dependent oxidation of NADH [34], did not affect the flavin fluorescence of PMCs. This again testifies in favor of the fact that the enzymatic oxidation of NADH by ferricyanide occurs without participation of flavin. And this is not surprising, because ferricyanide has iron with a *d*-orbital extending more than 10 Å, i.e. can easily reach the NADH pyridine ring sitting in the active center.

Unlike the oxidation reaction of NADH in the respiratory chain, the NADH: ferricyanide-reductase reaction is not sensitive to rotenone (a specific inhibitor; concentration – 6 μM). Insensitivity to rotenone is due to the fact that ubiquinone does not take any part in the NADH: ferricyanide-reductase reaction. Similar data were obtained for pNTP and DCPIP.

Apparently, iron-sulfur clusters also do not participate in the transfer of electrons to ferricyanide, since bathophenanthroline (specific chelator on iron) did not inhibit this reaction, although it inhibited the MC respiration (Fig. 5.10).

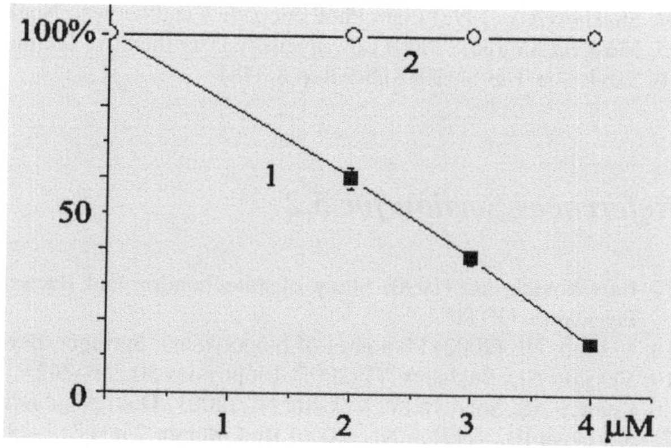

Fig. 5.10 Effect of bathophenanthroline on the oxidation of NADH in MCs. (1) NADH: CoQ reductase reaction, (2) NADH: ferricyanide-reductase reaction

5.3.1 Conclusion

The NADH-dehydrogenase complex is capable of carrying out two types of reactions: (1) reduction of ubiquinone or its analogues through flavin or iron-sulfur clusters and (2) reduction of artificial acceptors without participation of flavin and ubiquinone and may be without iron-sulfur clusters. The first reaction is rotenone-sensitive, and the second one is not. Rotenone-sensitive and rotenone-insensitive pathways of NADH oxidation are considered in [34]. The results presented were published in [40].

References

References Section for 5.1

1. Vekshin NL, Mironov GP (1982) Biophysics 27(3):537–539
2. Vekshin NL, Mironov GP (1982) Abstracts of the All-Union Biophysical Congress. V.1. ONTI, Moscow
3. Vekshin NL (1983) Physico-chemical bases of cell functioning. NCBI, Pushchino, pp 82–86
4. Vekshin NL (2002) Photonics of biopolymers. Springer, Berlin/London
5. Konev SV, Rudenok AN (1992) Biophysics 37(5):939–941
6. Ninnemann H, Strasser RJ, Butler WL (1977) Photochem Photobiol 26(1):41–47
7. Vekshin NL (1991) Biochem Int 25(4):603–611
8. Racker E, Stoeckenius W (1974) J Biol Chem 249:662–663
9. Passarella S et al (1984) FEBS Lett 175(1):95–99
10. Nedelina OS et al (1983) Biophysics 28(2):341–342
11. Dmitriev LF, Ivanova MV, Ivanov II (1990) Biol Membr 7(9):961–965
12. Stitt M, Lilley RMC, Heldt HW (1982) Plant Physiol 70:971–977
13. Ronzhina OA, Kravchenko EE (1988) In: Voznesensky VL (ed) Relationship of photosynthesis and respiration. TSU, Tomsk, pp 48–50

14. Shakhov AA (1993) Plant Photoenergetics and harvest. Nauka, Moscow
15. Moskvin SV (ed) (2000) Low-intensity laser therapy. Technique, Moscow, pp 71–94
16. Sanderson T et al (2018) Sci Rep 8:3481

References Section for 5.2

17. Belyakovich AG (1990) Study of mitochondria and Bacteria using pNTV Tetrazolium salt. Pushchino, ONTI
18. Vekshin NL (2002) Photonics of biopolymers. Springer, Berlin/London
19. Vekshin NL, Sukharev VI (2005) Biophysics 50:236–242
20. Lomtev AS, Sharova IV, Vekshin NL (2001) Dokl Akad Sci USSR 376:114–116
21. Sokolova IB, Vekshin NL (2006) Biol Membr 23(3):217–222
22. Sharova IV, Vekshin NL (2004) Biophysics 49:814–821
23. Panov AV (2015) Practical mitochondriology. Nauka, Novosibirsk
24. Mosolova IM, Gorskaya IA, Scholz KF, Kotelnikova AV (1975) Methods of modern biochemistry. Nauka, Moscow, pp 45–47
25. Vekshin NL, Sokolova IB (2006) Biophysics 51:288–296

References Section for 5.3

26. Brandt U (1997) BBA 1318:79–91
27. Ragan CJ (1985) Curr Top Bioenerg 15:1–36
28. Vekshin NL, Sharova IV, Sukharev VI (1999) In: Ghisla S (ed) Flavins and flavoproteins. Elsevier, Berlin, pp 635–638
29. Hatefi Y (1985) Annu Rev Biochem 54:1015–1069
30. Dooijewaard G, Slater E et al (1978) BBA 503:405–424
31. Walker JE et al (1995) Meth Enzym 260:14–34
32. Sled VD, Vinogradov AD (1993) BBA 1143:199–203
33. Vinogradov AD (1998) BBA 1364:169–185
34. Sharova IV, Vekshin NL (2000) Mitochondria, cells and reactive oxygen species. ONTI, Pushchino
35. Belyakovich AG (1990) Study of mitochondria and bacteria using pNTP tetrazolium salt. ONTI, Pushchino
36. Vekshin NL (1988) Biol Sci 4:107–111
37. Vekshin NL (2002) Photonics of biopolymers. Springer, Berlin/Heidelberg
38. Vekshin N (1991) Molek Biol 25:54–59
39. Dmitriev IS, Semenov SG (1980) Quantum chemistry. Atomizdat, Moscow
40. Lomtev AS, Sharova IV, Vekshin NL (2001) Dokl Acad Sci 376(1):114–116

Chapter 6
Red-Ox Reactions with NADH Dehydrogenase

6.1 Lag-Period and Pseudo-Oscillations in NADH Red-Ox Reactions with Dichlor-Phenolindophenol

Spectral properties of the DCPIP/DCPIP-H_2 pair are very convenient for the photometric determination of enzymatic activity of dehydrogenases: DCPIP is blue, and DCPIP-H_2 is colorless. Previously, DCPIP was used mainly in measuring the activity of succinate dehydrogenase [1]. DCPIP is a chemical analog of natural quinones.

When NADH was added to suspension of MCs (from bovine heart), a recovery of DCPIP to DCPIP-H_2 was observed, accompanied by discoloration of the absorption band at 600 nm, and then a spontaneous return of the color to the initial level was detected [2]. The first reaction is carried out by NADH dehydrogenase, and the second one is the oxidation of the reduced dye by oxygen on cytochrome oxidase. Moreover, DCPIP-H_2 is non-autooxidable. It oxidizes rapidly if cytochrome oxidase is not blocked [1].

The first reaction is about twice as fast as the second. One might think that just the difference in speeds that causes a certain separation of the two processes in time. However, it turned out that the matter is not only and not even so much. It was interesting to find out the conditions for an appearance of a lag period between the reactions of NADH-induced reduction and oxidation of the dye in mitochondrial shadows obtained in a hypotonic environment from MCs from bovine heart or rat liver. Shadows of MCs are not as muddy as the native MCs, so it is easier to detect by usual photometry. In addition, their membranes are highly permeable.

MCs were isolated from a bovine heart or rat liver based on standard methods [3, 4]. To measure the activity of NADH dehydrogenase, a suspension of cardiac or hepatic MCs (0.1–0.25 mg/ml) was added to the hypotonic incubation medium (20 mM Tris-phosphate buffer, pH 7.5–8.0) with 50–150 μM DCPIP and 150–200 μM NADH. Measurements were made in 0.5- or 1-cm quartz cuvettes as detection of optical density in the absorption region of DCPIP at 600 nm or NADH at

N. Vekshin, *Biophysics of Mitochondria*,
https://doi.org/10.1007/978-3-030-33853-4_6

340 (or 360 nm) in a special compartment for turbid samples of the M-40 spectrophotometer. Oxidation of DCPIP-H_2 to DCPIP was recorded by increasing in optical density at 600 nm. Oxygen consumption was measured polarographically [5]. The protein concentration was determined by the UV method [6]. MC shadows, painted by the dye (DCPIP or Janus green) were observed with an optical microscope.

When NADH is added to suspension of MC shadows (in a hypotonic medium), a rapid recovery of DCPIP is observed, accompanied by discoloration at 600 nm. The DCPIP-activated NADH oxidation (accompanied with dye reduction) is not sensitive to rotenone or amytal, as well as antimycin and azide. Rotenone-insensitive (RI) oxidation of NADH by mitochondrial NADH dehydrogenase, induced by artificial electron acceptors, is an order of magnitude faster than rotenone-sensitive (RS) oxidation [7]. The reason for this is the shunting of electron transfer bypassing flavin and ubiquinone [2]. After partial bleaching of DCPIP, a spontaneous color return almost to the initial level occurs. A similar pattern was observed for hepatic mitochondrial shadows. The first reaction (rapid discoloration) is carried out by NADH dehydrogenase, and the second reaction, the slower one, is the oxidation of DCPIP-H_2 by oxygen on cytochrome oxidase.

When measurements were taken in a cuvette without mixing, repeated spontaneous alternation of the first and second reactions was observed, even visually observed by the appearance of a blue color from above (oxygen from the air goes into solution) of the cuvette and the colored area spreads down with subsequent discoloration. Such "pseudo-autowave oscillations" (Fig. 6.1) depend on the non-uniformity of aeration of suspension in the cuvette. This effect can serve like the simplest model of a "wave" cascade of oxygen-dependent red-ox reactions in a living cell. When the experiment was conducting in anaerobic conditions, repeated "fluctuations" were not observed.

Pre-incubation of the MC shadows with succinate did not affect the rate of accumulation of DCPIP-H_2 during the oxidation of NADH. However, incubation with succinate directly during the experiment in a closed cuvette (or in an open, but without stirring) led to the cessation of oxidation of DCPIP-H_2 (Table 6.1).

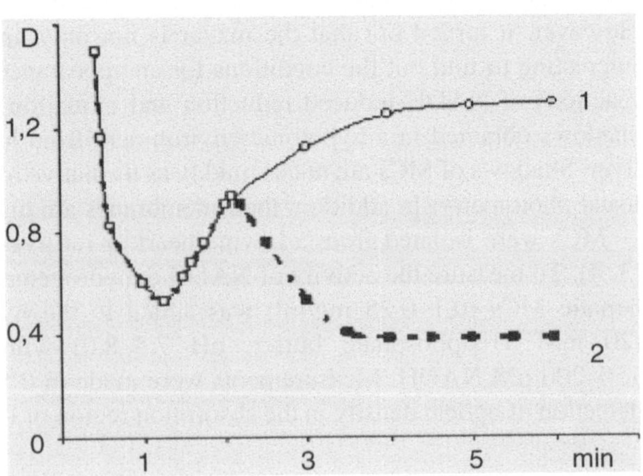

Fig. 6.1 Kinetics of DCPIP/DCPIP-H_2 reactions at 600 nm during oxidation of NADH in suspension of cardiac mitochondrial shadows. DCPIP – 100 μM, NADH – 160 μM, MC shadows – 0.1 mg/ml (in 20 mM Tris-HCl, pH = 8). 1 – with stirring, 2 – without

Table 6.1 The rate of the NADH-dependent reduction of DCPIP (V1) and the subsequent oxidation of DCPIP-H_2 (V2) in suspension of mitochondrial shadows under various conditions

Conditions	V1, %	V2, %	Lag phase, min
Control	100	100	~ 0.2
NAD, 600 μM	90	95	~ 0.2
Amital, 1 mM	~ 100	90	2
Rotenone, 5 μM	~ 100	90	3
Antimycin, 2 μM	~ 100	90	4
Azide, 5 mM	~ 100	63	8
Succinate, 5 mM	100	100	~ 0.2
Succinate, 5 mM + anaerobiosis	100	0	–
30 min warming up, 50 °C	230	~ 80	3
3 h warm-up, 50 °C	~ 50	0	0
30 min warming up, 90 °C	0	0	–

Succinate, as is known, restores the respiratory chain and stimulates intensive oxygen consumption, which creates anaerobic conditions in the cuvette. With sufficient oxygen supply (intensive mixing in an open cuvette), the addition of succinate did not lead to the disappearance of the second reaction, but contributed to a greater oxidation depth of the dye.

In the control experiments, after non-enzymatic reduction of DFIPP by ascorbate, the dye discoloration was also observed (data not shown). The addition of mitochondrial shadows led to the rapid oxidation of DCPIP-H_2. This reaction, which proceeds on cytochrome oxidase, was sufficiently suppressed by sodium azide (Table 6.1).

All these data indicate the oxygen-independent nature of the first reaction and oxygen-dependent – the second. During the first reaction, no noticeable consumption of oxygen was observed, and during the second, it was observed (polarographic data are not given). All subsequent experiments were conducted under aerobic conditions in open cuvette with vigorous stirring. Under these conditions, the first and second reactions occur only once.

The addition of NAD had a negligible effect on the rate of both reactions and on the lag period (Table 6.1). The lack of noticeable competition between NADH and NAD during the NADH: DCPIP-oxidoreductase reaction means that under shunting conditions the NADH dehydrogenase has two separate binding centers – for NADH and for NAD. The addition of DTT stimulated the first reaction and irreversibly inhibited the second. Acceleration of the first reaction may be associated with non-enzymatic reduction of NAD by DTT to NADH. Inhibition of the second reaction is a consequence of the deactivation of cytochrome oxidase by DTT.

Heating the MC shadows at 50 °C for 30 min (with followed cooling to room temperature) led to a strong stimulation of the first reaction and some inhibition of the second (Table 6.1). The acceleration of the first reaction can be explained by an increase in the availability of binding sites for DCPIP or NADH. The decrease in the rate of the second reaction is associated with damage of cytochrome oxidase.

Fig. 6.2 Effect of respiratory chain inhibitors on DCPIP/DCPIP-H_2 reactions. DCPIP – 100 μM, NADH – 160 μM, hepatic liver MC shadows – 0.1 mg/ ml (in 20 mM Tris-HCl, pH = 8). 1 – without inhibitors, 2 – with amytal, 3 – with rotenone, 4 – with azide

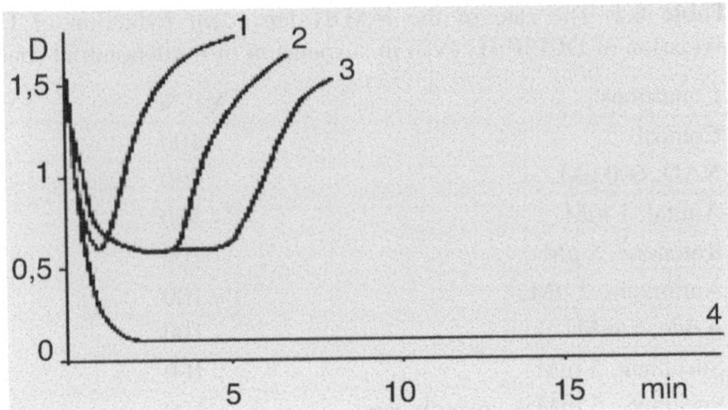

Heating at 50 °C for 3 h led to a sharp slowdown of the first reaction and complete suppression of the second. An increase in temperature to 90° C (within 30 min) led to a complete inhibition of both reactions due to a total inactivation of all enzymes.

Amital, rotenone, antimycin and azide sharply extended (in many times) the lag-period between two reactions, up to 8 min, and, in the case of a large concentration of azide, the second reaction did not occur at all (Fig. 6.2).

In this case, the first three substances had little effect on the oxidation rate of DCPIP-H_2. This means that although ubiquinone and cytochromes b and c do not directly participate in the reduction-oxidation reactions of the dye, here the lag period is caused by blocking of the respiratory chain between NADH-dehydrogenase and cytochrome oxidase. The lock is temporary. This means that during the lag period some kind of "azide-resistant" path is activated. Azide not only dramatically lengthened the lag-period, up to complete inhibition, but also markedly deepened the level of dye recovery. The strong lengthening of the lag-period by azide, as well as the decrease in the rate of the second reaction, indicate that cytochrome oxidase was suppressed. In other words, after inhibition of the respiratory chain with amital, rotenone, antimycin, or azide, all the available DCPIP-H_2 molecules during several minutes are not capable of donating electrons through cytochrome oxidase to oxygen, and then the "block" is spontaneously removed. Over time, the activation of the "azide-resistant pathway" of the oxidation of DCPIP-H_2 appears. There is a slow diffusion of DCPIP-H_2 molecules through the inner membrane, as a result of which cytochrome oxidase begins to oxidize the DCPIP-H_2 (Fig. 6.3).

To answer the question of how heavily the used dye binds to mitochondrial shadows, they were passed at the presence of DCPIP through 0.22-micron Millipore filter. At the same time, 60% of the dye settled on the filter together with the shadows (DCPIP without shadows was not absorbed by this filter). Optical microscopy of MCs was also performed in the presence of DCPIP. It was confirmed that DCPIP stains MC shadows very well.

In a number of works, it was previously assumed that the NADH oxidation pathway is carried out not through NADH dehydrogenase of respiratory chain, but through another enzyme located in the outer membrane (b5-shunt) [8–13] or on the outer side of the inner membrane [14] ("external path"). However, it is shown that

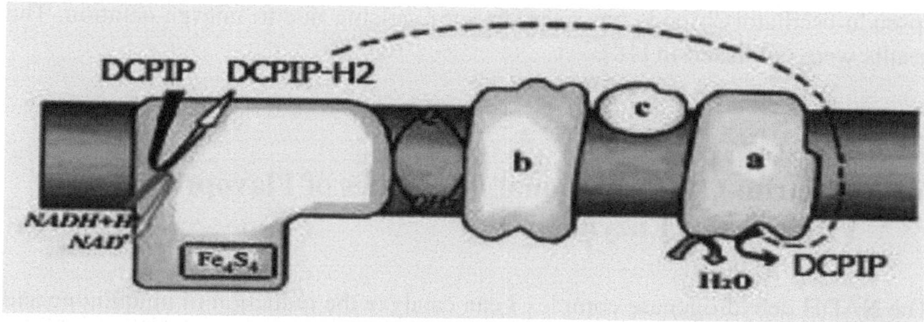

Fig. 6.3 Electron shunting from NADH dehydrogenase to DCPIP and oxidation of DCPIP-H_2 on cytochrome oxidase

there is no any "external" NADH dehydrogenase in MCs and the RI oxidation of NADH occurs with the participation of the damaged NADH dehydrogenase of the inner membranc [2]. The obtained data confirmed that this shunt is not on the outer membrane, but on fragments of the NADH dehydrogenase [7]. Interestingly, most preparations of isolated NADH-dehydrogenase are not sensitive to rotenone or low sensitive [15, 16].

On the basis of the data obtained, it can be concluded that the shunting of electrons from NADH dehydrogenase through DCPIP to cytochrome oxidase has a slow stage, which determines the lag period. The duration of the lag-period depends on the functional state of respiratory chain.

Since DCPIP is similar to natural quinones in its chemical nature and red-ox properties, it is possible that under certain conditions membrane ubiquinone can (by a similar way) bypass electron transfer from NADH dehydrogenase to cytochrome oxidase, bypassing other components of the chain. Ubiquinone-dependent red-ox reactions in cyanide-resistant respiration of MCs were investigated, for example, in [17].

6.1.1 Conclusion

The recovery of the oxidation of NADH occurs by shunting electron transfer in the initial part of complex I of the respiratory chain. A lag-period (not less than 20 s) is need for beginning to oxidize DCPIP-H_2 molecules on cytochrome oxidase. DCPIP-H_2 is not initially capable of entering into the indicated reaction, but some slow stage is required, probably – diffusion of DCPIP-H_2 through the inner membrane. The fact that the inhibitors of the respiratory chain – amital, rotenone, antimycin and azide very significantly, up to 2–8 min, prolonged the lag-period, testifies in favor of the activation of the "azide-resistant pathway". The studied processes can serve as a model for shunting electron transfer from NADH dehydrogenase to cytochrome oxidase through ubiquinone, bypassing other components of the respiratory chain. A "pseudo-oscillatory" mode has been detected, arising due to uneven aeration. The results were published in [18].

6.2 Electron-Conformational Properties of Flavoprotein Fragment of Enzyme

The NADH dehydrogenase complex I can catalyze the reduction of ubiquinone and proton transport coupled with electron transport [19, 20]. The complex looks like a "boot" with a size of 100 Å [21]. When exposed to chaotropic agents, the complex is divided into a water-insoluble membrane part and a water-soluble peripheral part. Peripheral part shows high catalytic activity in the presence of artificial electron acceptors [22, 23]. The peripheral part, depending on the method of isolation, may contain a different number of subunits, iron-sulfur clusters and flavins.

Almost all methods to obtain the NADH-dehydrogenase complex or its active fragments [20, 21, 31, 131, 134, 136–138] are heavy laborious. Most of them give a low yield of active enzyme, poorly reproducible and often do not allow avoid the loss of flavin and iron. In addition, procedures such as sonication or high concentrations of detergents and chaotropic agents can significantly affect the properties of the enzyme.

Enzyme preparations are generally insensitive to rotenone and have a high NADH: ferricyanide-reductase activity. During the isolation of complex I, the degree of its nativeness is estimated by NADH: ferricyanide reductase activity. However, high activity with artificial acceptors is in fact not an indication of the native nature of the isolated enzyme. When the enzyme is released, it usually turns into a non-native conformation.

It was interesting to study the generalized electron-conformational properties of the peripheral fragment of NADH dehydrogenase using fluorescence spectroscopy — time-resolved pulsed, phase-modulation, polarization, and correlation confocal. Tryptophan residues and flavin of the protein itself were used as fluorophores.

MC membranes were isolated from the heart muscle of a bull according to [30]. The selection of the fraction of the peripheral fragment of NADH dehydrogenase from MCs was carried out according to [22, 31] with modifications [29].

The resulting fraction contained iron-sulfur clusters and flavins, namely: FMN – 70%, riboflavin – 15%, FAD – 15% [24, 29]. The presence of 15% FAD suggests that FMN and riboflavin in the preparation may be the result of a rapid hydrolysis of FAD, catalyzed by the nearby iron-sulfur cluster. It is interesting, that all three flavins were in a binding form, not free. The peripheral fragment, consisted of 8 subunits of NADH-dehydrogenase, was at least 85% of all proteins in the fraction. The fraction had a high activity relating to oxidation of NADH by artificial electron acceptors—ferricyanide, DCPIP, and pNTV [25, 29].

Polarized tryptophan emission of the fraction of the peripheral fragment of NADH-dehydrogenase was measured in 20 mM Tris or phosphate buffer (pH 7.5) on “SLM-4800” with 300 nm edge excitation and detection at 340 nm, and polarized flavin emission – at 460 and 520 nm, respectively. The lifetime τ was measured by the phase method at 30 MHz. At the experiments, mirror cuvettes were used, which allowed to enhance the fluorescent signal [27].

Substituting the lifetime τ and the measured degree of polarization P into the Levshin-Perrin equation [27] allow us to estimate the magnitude of the rotational mobility (angle θ) of the fluorescent protein and calculate its volume (V):

$$1/P - 1/3 = (1/P_0 - 1/3)\,(1 + RT\tau/\eta V) = (1/P_0 - 1/3)\big(2/\big(3\cos^2\theta - 1\big)\big)$$

Here P_0 is the limit polarization ($P_0 = 0.42$ for flavin [24] and the same for tryptophan [32]), R is the universal constant, T is the absolute temperature 293K, η is the solution viscosity 0.01 poise.

The fluorescence decay kinetics [26] of tryptophan residues was measured at 20 °C (excitation 300 nm, pulse frequency 951.2 kHz, temporal resolution 20 picoseconds; the 348 nm interference filter was used to isolate tryptophan emission).

Correlation microscopy was performed on ConfoCor. Laser radiation of 488 nm with a power of ~ 20 mW passed through a weakening filter and was focused in a volume of 0.1–0.4 femtoliter [28]. Concentrations of particles were: protein – 2.8 nM, PMCs – 0.6 nM, flavins – 89 nM, FMN in DDS – 8.2 nM. The diameter (d) of particles was determined by the equation [28]:

$$d = kT/3\pi D\eta$$

Here k is the Boltzmann constant, D is the coefficient of translational diffusion, $D = \omega^2/4\tau_d$, where τ_d is the time required for the particle to diffuse over a distance ω (the cross section of the ellipsoid equal to the square root of πr2).

6.2.1 *Polarization and Phase Modulation Spectroscopy*

The phase lifetime of flavins in the peripheral fragment is about 4.5 ns (Table 6.2).

A high degree of polarization of 0.3 means that the proper rotation of the flavin can be neglected, and then when substituting the lifetime and polarization, it is possible to calculate the effective size of the protein globule using the Levshin-Perrin equation. Similarly, you can do by measuring the lifetime and polarization of tryptophan emission. The diameter of the protein, found in both ways, is not less than 44 Å (Table 6.2).

The quantum yield of flavin fluorescence in the peripheral fragment is very low due to the rapid deactivation of the excited flavin by the cluster Fe_4S_4 [24, 25]. There

Table 6.2 The degree of polarization and phase life-time of tryptophan residues and flavins in the peripheral fragment of NADH dehydrogenase and the size of the protein globule calculated by the Levshin-Perrin equation

	Trp	Flavin
Spectral maximum, nm	340	525
Life-time, ns	2	4.5
Polarization	0.36	0.3
Rotation angle, degree	18	26
Global volume, Å^3	46,600	43,700
Diameter of the globule, Å	45	44

is a strong electron-vibrational bond between flavin and iron. The overlapping of electronic clouds of flavin and iron-sulfur cluster can provide fast electron transfer between them.

6.2.2 *Tryptophan Fluorescence Decay*

Decay of the tryptophan fluorescence of the peripheral fragment is strongly non-exponential (Table 6.3). The main contribution belongs by short-lived components, i.e. tryptophan residues are strongly quenched.

The decay kinetics of tryptophan fluorescence has three components: 0.14, 1.05, and 3.27 ns with amplitudes of 33.6, 38.6, and 24.4%, respectively. The most powerful quenchers in proteins are iron-sulfur clusters, which reduce the lifetime of tryptophan residues to 0.1–1 ns [26]. The peripheral fragment of NADH-dehydrogenase contains a number of FeS clusters [22, 23]. The two decay components – the shortest-living ones – belong to those tryptophan residues, which are located near the FeS clusters, in the radius of the electronic orbital of iron – about 10 Å.

The lifetime τ of each tryptophan residue depends on the microenvironment, and the amplitude depends only on the amount of tryptophan residues with a certain τ [27]. Amplitudes are proportional to the number of the residues in a given environment. This means that the 0.14- and 1.05-ns components, which together make up 72% of the amplitude, belong to those tryptophan residues that are deactivated by iron. Most of the tryptophan residues in this protein are placed at the sphere of iron orbital. Apparently, such electronic overlapping of tryptophan residues with iron-sulfur clusters facilitates electron transfer through the protein structure.

The NADH: ferricyanide-reductase activity was completely lost after 2 h of heating at 40°C. Meanwhile, as can be seen from Table 6.3, no any changes in the tryptophan fluorescence decay are observed after the sample was heated. This indicates the absence of significant conformational changes after warming up at 40°C, as well as the fact that the cause of the temperature inactivation of the enzyme cannot be the loss of iron.

Table 6.3 Parameters of the decay kinetics of tryptophan fluorescence of the peripheral fragment of NADH-dehydrogenase before and after heating at 40 °C

Parameters		Protein		Inactivated protein	
τ_1 (ns)	α_1 (%)	0.14	33.6	0.12	35.1
τ_2 (ns)	α_2 (%)	1.05	38.6	0.99	37.5
τ_3 (ns)	α_3 (%)	3.27	24.4	3.28	24
r_i		0.248		0.246	

Note: τ_1, τ_2, τ_3 – the first, second and third components of fluorescence decay, α_1, α_2, α_3 – their amplitudes; r_i is the initial anisotropy. Inactivation of the protein was carried out by heating at 40 °C for 2 h (followed by cooling to 20 °C)

Table 6.4 Correlation microscopy of flavin fluorescence fluctuations and calculated diameters of particles

Sample	τ_d (μs)	d *(Å)*
FMN in water	*59 ± 7*	*15 ± 2*
FMN with SDS	*295 ± 4*	*75 ± 1*
Enzyme	*360 ± 5**	*72 ± 2*
PMC	*9000 ± 100*	*2285 ± 25*

Notes: τ_d is the diffusion time (it was measured in an illuminated volume of 1.1 femtoliter), d – the particle diameter, PMCs – protomitochondria obtained by filtering of MC suspension through a filter with a pore diameter of 0.22 micron

6.2.3 *Correlation Confocal Microscopy of Flavins*

Correlation spectroscopy allows one to estimate the sizes of diffusing particles by detection of fluctuations of the fluorescence intensity inside a microscopic illuminated volume [28]. As can be seen from Table 6.4, the free FMN has a diffusion diameter of about 15 Å (a solvent shell makes a significant contribution to this value).

Free FMN can be quickly incorporated from an aqueous solution into SDS micelles. The diameter of the particles SDS with flavin is 75 Å (Table 6.4). This value is consistent with those values that were obtained using correlation microscopy in [28] for SDS micelles with fluorescently labeled phospholipids.

The effective diameter of the diffusing protein globule of the enzyme is 72 Å. This value is higher than that obtained for the rotating globule by the Levshin – Perrin eq. (44 Å). The difference may be due to: (a) the difference in the diffusion and rotational diameters of the anisotropic protein structure; (b) underestimated values of the rotational diameter due to the neglect of the rotational movement of the fluorophore inside the protein globule; c) inaccurate determination of the diffusion coefficient.

PMC particles, containing NADH-dehydrogenase and other components of the respiratory chain [29], have a diffusion diameter of 2280 Å (Table 6.4), which coincides with the pore diameter of the filter used in the preparation of particles—0.22 microns.

6.2.4 *Conclusion*

The effective rotational diameter of peripheral fragment is not less than 44 Å, and the translational diameter is 72 Å. The tryptophan and flavin emission of the fragment is strongly quenched by the iron clusters, which indicates a strong electron-vibrational interaction of iron with tryptophan residues and flavin. It is possible that the overlapping of the electron orbitals of iron-sulfur clusters, tryptophans and flavin facilitates electron transfer through the protein structure. The results were published in [33].

6.3 Resonant Energy Transfer from Tryptophans to NADH

During the transitions of MC from one functional state to another, caused by the addition of ADP or respiratory substrates, small (5–10%) changes in the intensity of protein fluorescence were detected [34–38]. These changes were attributed to conformational transitions of membrane proteins [34, 35, 37]. On the other hand, it was discovered [39] that in response to the addition of ADP or succinate, changes in NADH fluorescence occur [40]. Changes in the intensity of the MC protein fluorescence are mirrored to changes in NADH fluorescence [36]. Since the spectrum of tryptophan emission overlaps with the absorption spectrum of NADH, an idea emerged about the possibility of resonant energy transfer from tryptophan residues of the NADH dehydrogenase to the bound NADH [35, 36]. It was suspected that changes in protein fluorescence are not caused by conformational transitions, but are a reflection of changes in the efficiency of resonant energy transfer from tryptophans to NADH [36]. However, there was no evidence.

MCs were isolated from rat liver as in [41]. The MC suspension (70 mg of protein in 1 ml) was kept on ice in medium containing 300 mM sucrose, 20 mM Tris-HCl, pH 7.6. Fluorescence was measured on Perkin-Elmer MPF-44B at 20 °C. The excitation spectra were recorded in the "Ratio" mode using a 0.1-cm cuvette. The measurements were carried out in hypotonic (5 mM Tris-HCl, 4 mM KH_2PO_4, pH 7.5) or isotonic (5 mM Tris-HCl, 4 mM KH_2PO_4, 50 mM KCl, 150 mM sucrose, pH 7.5) medium. In contrast to [36], a low concentration of MCs was used (0.05 mg of protein per ml in a 1-cm cuvette and 0.3 mg of protein/ml in a 0.1-cm cuvette), which made it possible to avoid screening, reabsorption and multiple light scattering. The optical density at 286 nm was 0.084–0.14 [42]. The lifetime τ was determined by the phase method [43] at 30 MHz.

For the donor-acceptor pair tryptophan/NADH, the theoretical "Förster radius" (the distance at which the probability of energy transfer and the probability of deactivation are the same [43, 44]) reaches 30 Å [35, 38]. Actually, in proteins, the effective energy transfer is usually carried out at distances comparable to the size of donor and acceptor chromophores ~ 10 Å [38, 44]. For example, in alcohol dehydrogenase, there is an almost 100% energy transfer from Trp-314 to NADH,

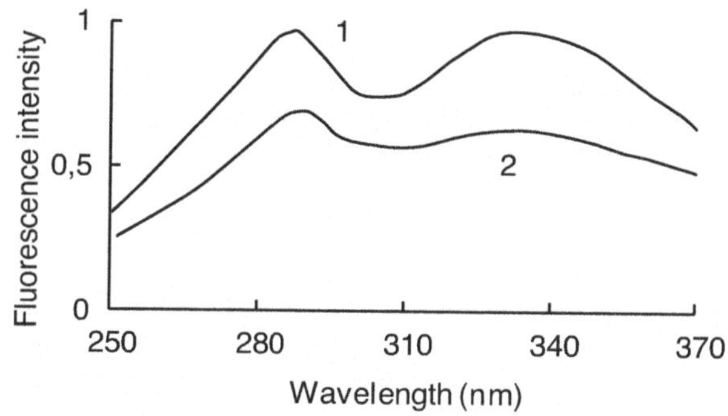

Fig. 6.4 Excitation spectra of mitochondria (0.3 mg per ml; emission at 440 nm) are initial (1) and after 2 min. After adding 80 μM $CaCl_2$ (2) in isotonic medium

less than 17 Å between the centers of the indole and pyridine chromophores, but no transfer from Trp-15 due to too long distance of 27 Å [38]. In yeast alcohol dehydrogenase [45] and hepatic isocitrate dehydrogenase [46], pyridine nucleotides are bound in the immediate vicinity of tryptophans.

NADH binding sites in MC are NADH dependent dehydrogenases. Unlike free NADH, which has a maximum fluorescence in water at 460–470 nm [5], the fluorescence spectrum of endogenous pyridine nucleotides NADH and NADPH in MCs has a maximum at 440–450 nm, which is due to protein binding [39].

A reliable sign of resonant energy transfer is the appearance of a donor "absorption band" in the excitation spectrum of a fluorescent acceptor [44]. Two maxima are observed in the excitation spectrum of mitochondrial NADH: the tryptophan band in the region of 286 nm and the NADH band at 335 nm of almost the same intensity (Fig. 6.4). This spectrum is similar to that of energy transfer from tryptophan to NADH in alcohol dehydrogenase [38].

The ratio between the number of donor tryptophans and NADH acceptor molecules can be estimated from the intensity of the tryptophan band in the NADH excitation spectrum:

$$[d]/[a] = F_{sa}\varepsilon_a/F_a\varepsilon_d Q_m$$

Here [d] and [a] are the number of donors and acceptors, F_{sa} and F_a are the fluorescence intensities of NADH upon excitation at the donor (286 nm) and acceptor (340 nm) absorption bands, ε_d and ε_a are the molar extinction coefficients of the donor and acceptor, Q_m is the energy transfer efficiency [47]. In proteins, tryptophans have $\varepsilon_d = 5600\ M^{-1}\ cm^{-1}$ [35, 38], which is only slightly less than that of NADH at 340 nm – 6200 $M^{-1}\ cm^{-1}$ [38]. Judging by the intensities of the bands in the excitation spectrum (Fig. 6.4) and assuming $Q_m = 1$, we can conclude that only one tryptophan residue (on average) serves for one NADH acceptor molecule (at 286 nm, a small contribution of NADH itself was taken into account).

When MCs are placed in hypotonic medium, a spontaneous increase in their tryptophan fluorescence by 12% is observed (Fig. 6.5a). At the same time, a decrease in endogenous NADH fluorescence on 50% is observed. In a hypotonic environment, as is known [36, 48], MCs are "swelling": endogenous calcium phosphate is released,

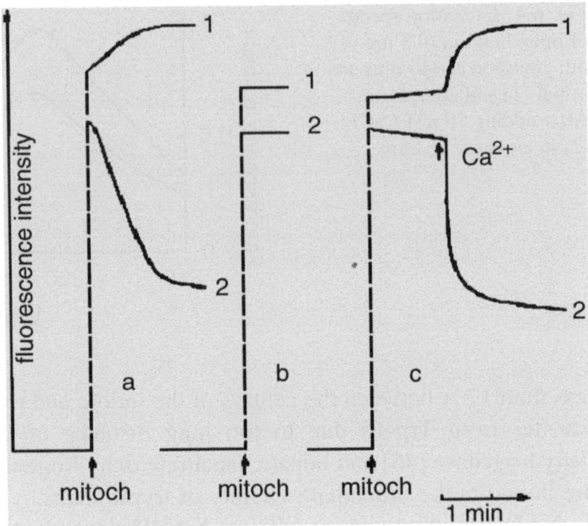

Fig. 6.5 Fluorescence of MCs at 330 nm with excitation at 286 nm (1) and at 440 nm with excitation at 340 nm (2): **a** – is hypotonic medium; **b** – the same medium +0.5 mM EGTA; **c** – isotonic medium. The protein concentration is 0.05 mg/ml. $CaCl_2$–20 μM

phospholipases are activated, and oxidation of pyridine nucleotides is accelerated. In this case, there is a small oxygen consumption caused by the oxidation of endogenous substrates and NADH. The induction of oxidation of added NADH under osmotic shock of MC is well known [48]. The fall in fluorescence of NADH is caused by oxidation to NAD, which, unlike NADH, has no absorption at 340 nm and therefore does not emit at 440 nm. After the second minute, the both fluorescent kinetics reach a plateau corresponding to the equilibrium between NADH and NAD. Since the addition of EGTA – complexon for calcium – prevents the changes (Fig. 6.5b), this means that calcium ions released from the MC matrix are the inducer of NADH oxidation (in the hypotonic medium). In isotonic environment, spontaneous fluorescent changes are very slow, but sharply stimulated by the addition of calcium ions (Fig. 6.5c). Similar, but more complex changes in the fluorescence of MC in isotonic medium in response to the addition of calcium were observed in [36]. All the results obtained are consistent with known data on calcium-induced oxidation of NADH, suppressed by EGTA [49].

Thus, the enhancement of tryptophan fluorescence is caused not by conformational transitions per se or by decrease in the efficiency of energy transfer to NADH due to conformational changes, but – by the oxidation of the energy acceptor NADH to NAD. This is also indicated by the fact that after the kinetics reach the plateau the ratio between the intensities of the bands in excitation spectrum does not differ much from the initial one (Fig. 6.4). Constancy of a ratio is characteristic for invariable efficiency of energy transfer.

Lifetime τ of free NADH in water is ~ 0.4 ns [38]. Lifetime of the mitochondrial NADH is 1.8 ns. This value coincides with that for NADH in alcohol dehydrogenase [38]. Probably the almost entire mitochondrial NADH is bound to proteins. When the fluorescence intensity of NADH decreased, the τ value did not change. This means that the decrease in the fluorescence intensity of NADH is a result of its oxidation, but not the result of the dynamic quenching or desorption from proteins into water.

6.3.1 Conclusion

The resonant energy transfer from tryptophan residues to bound NADH in MCs has been confirmed. On average, only one tryptophan residue is an energy donor for one NADH acceptor molecule. The increase in the fluorescence of mitochondrial proteins, observed in hypotonic medium and in response to the addition of calcium, is due to the oxidation of NADH, but not to the conformational changes of proteins.

6.4 Loss of Flavin from Enzyme

When complex I damaged, it lost the NADH: ubiquinone reductase activity [50–53]. High probability of damage is due to several reasons. First, this complex in the respiratory chain is the first key. Secondly, it is extremely large. Thirdly, it contains 6–8 labile iron-sulfur clusters and a weakly bound flavin. Fourth, its functioning is associated with proton transfer and ATP synthesis. Therefore, the slightest damage leads to an avalanche of disorders, to mitochondrial diseases [54–58].

The peripheral flavoprotein fragment of complex I consist of three main subunits: 51, 24, and 9 kDa. The heavy 51-kDa subunit contains flavin (FMN or, possibly, FAD) and is responsible for the binding of NADH [52, 53]. The fragment catalyzes the rapid rotenone-insensitive oxidation of NADH by artificial acceptors, including ferricyanide. Neither ubiquinone nor flavin participates in the NADH: ferricyanide reductase reaction [59–61]. When the fragment is divided into individual components, flavin goes into solution in the form of FMN, and the obtained fractions (51 kDa and 24 + 9 kDa) become catalytically inactive [62].

Complex I play an important role in the formation of superoxide [63, 64] and lipid peroxidation [65], leading to membrane damage and, as a result, to dysfunction. Disturbances of complex I is a characteristic of aging [56]. In MCs of aging organisms, lipid peroxides accumulates, therefore, proteins and DNA are damaged. Neuromuscular systems are involved in the pathological process, requiring a lot of energy. If any damage appears at the level of complex I, the ATP ceases to be produced there [57, 66]. That is why encephalomyopathy occurs. Damage of complex I in the brain and muscle cells plays a crucial role in the development of Parkinson's disease and others disorders [55, 66–73].

Disruptions in the operation of complex I may occur under the action of various poisons and aggressive chemical substances. These include rotenoids, antibiotics, pesticides, barbiturates, etc. [74]. Such substances block the active centers of the enzyme or violate its conformation. There is evidence of inhibition of complex I by non-ionic detergents like Triton X-100 [75, 76].

The cause of a number of mitochondrial pathologies may be a loss of flavin from the active center of the enzyme. Flavin in the enzyme is the "weak link". The loss of flavin (in the form of FMN or riboflavin) can occur as its release into aqueous phase due to conformational rearrangements of the enzyme. Although flavin is located inside the protein globule, it can easily be lost, because bound not firmly. A loss of flavin during the isolation and purification of the enzyme is well-known [50, 51, 61]. The purified enzyme usually has a low NADH: ubiquinone reductase activity, but a high NADH: ferricyanide reductase activity [50–53].

It was interesting to find out under what conditions the flavin can leave the MC and how this process is stimulated or inhibited. Such factors as temperature, pH and detergents have been investigated [83]. As a model object the liver rat MCs were taken. Properties of MCs of brain, muscles and liver are differed not too much [77].

MCs from rat liver were isolated by standard method with modifications [59]. The protein concentration was determined by the UV method [78]. A suspension of diluted MCs (0.3 mg/ml) in an isotonic medium was divided into samples. Substrates, agents or detergents were added to samples. MCs were incubated for 2–4 h at room or other temperature, taking samples for subsequent measurements of a) released flavin (according to its fluorescence) and b) enzymatic activity.

The intensity and degree of polarization of flavin fluorescence was measured at baseline and after 2–4 h of incubation of the MC suspension. Experimental samples were compared with two controls (sample without additives, standing at room temperature; sample without additives, standing on ice). The measurements were carried out with diluted samples on "SLM-4800", $\lambda ex = 450$ nm (maximum of the flavin absorption band), $\lambda em = 525$ nm (emission maximum). The diluted fraction of MCs was placed in mirror cuvette, allowing the signal to be many-times amplified [79]. Since even a diluted suspension of MCs is optically scattering, the contribution of light scattering was subtracted from emission in all experiments. For this purpose, a glycogen solution was prepared with the same optical scattering density as samples. Then the light scattering of glycogen and samples were leveled, controlling this with the help of the "SLM-4800" at the same excitation and emission wavelengths (with slits of 1 nm). Then, the scattering spectrum of the glycogen solution was recorded at the same wavelengths (at $\lambda ex = 450$ nm) and under the same conditions as the sample fluorescence. Then the subtraction was made.

The phase and modulation lifetime of flavins of MCs was measured in mirror cuvettes on the "SLM-4800" at a modulation frequency of 30 MHz. The degree of polarization was measured on the same instrument, taking into account the contribution of light scattering according to [79].

To measure the activity of NADH dehydrogenase, the diluted fraction of MCs in 5 mM or 20 mM Tris-phosphate buffer (pH 7.5 at 20 °C) was placed in 1-cm quartz cuvettes into a Specord UV Vis spectrophotometer or M-40. NADH oxidation was determined by the decrease in optical density at 340 nm. The recovery of ferricyanide was also measured by the decrease in optical density at 410 nm.

For a long time, it was believed that changes in the fluorescence intensity of flavins during the functioning of MCs are associated exclusively with the processes of their oxidation-reduction and with the electron transfer through the respiratory chain. Although FMN-H_2 and FAD-H_2 do not fluoresce, unlike their oxidized forms – FMN and FAD, but other reasons for the change in flavin fluorescence are possible, in particular, the release of flavins from membranes into aqueous phase. Let consider this in details.

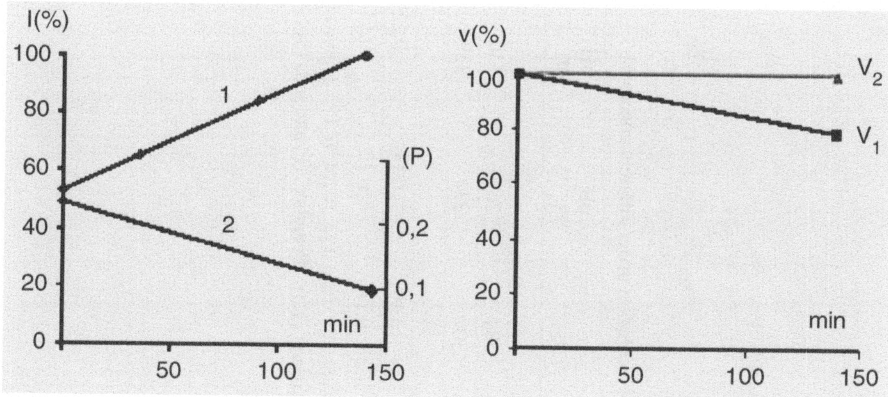

Fig. 6.6 Kinetics of spontaneous changes in the intensity and polarization of flavin fluorescence during incubation of mitochondria at 20 °C. Change in the rate of NADH: ubiquinone reductase reaction (V2) and NADH: ferricyanide reductase reaction (V1) when mitochondria were incubated at 20 °C

The fluorescence intensity of flavins in MCs is low (due to quenching by iron-sulfur clusters and light scattering on a single MC), and the degree of polarization is high (due to low rotational mobility and a short lifetime τ). The flavin fluorescence of MC is caused not only by FMN (or FAD) of complex I, but also by other FAD-containing proteins, contributing about 30% to the total intensity [60]. Judging by spin characteristics of cluster Fe_4S_4 [50] in complex I, it places next to the flavin, that can lead to quenching of the flavin fluorescence [79].

Flavins can spontaneously exit from MC. Free FMN (or riboflavin) in solution has a high fluorescence intensity and zero degree of polarization (P) [79]. When the MC suspension is incubated a long time, the intensity of the flavin fluorescence increases and P decreases (Fig. 6.6), i.e. flavins are released into aqueous phase. After 2 h of incubation, the intensity increases 1.5–1.8 times, and after 4 h – 3.3 times (Table 6.5).

For intact freshly isolated MCs, the initial P was 0.31. After 2 h of incubation, it decreased to 0.19 and after 4 h to 0.14 (Fig. 6.6, Table 6.5). This suggests that during the incubation of MCs, the rotational mobility of the flavins increases. This increase is due to the release of flavins (in the form of FMN and riboflavin) into aqueous phase. When MCs are incubated on the ice, the release of flavins is much slower than at room temperature (Table 6.5).

The fact that FMN or riboflavin is released from MC, and not FAD, was determined using the ethylene glycol method [80]. In addition, FAD and FMN can be distinguished by the lifetime of the excited state τ. In water, the τ of FMN is 4.5 ns, and FAD is 2.5 ns [79]. The lifetime of the flavin fluorescence of the MC was initially non-exponential (3 ns with phase and 5.4 ns with modulation), i.e. on average, 4.2 ns, which is close to τ of FMN.

Table 6.5 Changes in the intensity and polarization degree of mitochondrial flavin fluorescence after 2 and 4 h of incubation under different conditions

h	B	BI	NAD	SDS	So	Fa	Ko	Pe	pH 3	pH 10
Fluorescence intensity %										
2	155	148	98	184	147	334	257	184	269	130
4	334	175	171	360	275	496	451	307	374	189
Polarization										
2	0.19	0.31	0.31	0.11	0.14	0.1	0.11	0.19	0.14	0.22
4	0.14	0.16	0.28	0.13	0.16	0.12	0.1	0.11	0.12	0.13

Note: Incubation and all measurements were carried out at 20 °C, except for sample BI – buffer on ice. The initial intensity was taken as 100%. The degree of polarization at the initial moment (0 h) in the buffer (B) was 0.31. NADH and NAD – 10^{-4} M. SDS, "Fairy" (Fa), "Komet" (Ko), "Pemolux" (P) and soda (So) – 0.001%

As well known, enzymes can be stabilized by their substrates, reducing the dynamics of the structure. The spontaneous release of flavin was slowed down at incubation of MCs with NAD, which stabilized the active center of complex I (Table 6.5). In addition, NAD is able to suppress the formation of superoxide in complex I and thereby protect it from damage [81].

The loss of flavin from complex I accompanied by a decrease in the rate of NADH: ubiquinone-reductase reaction (Fig. 6.6). Moreover, flavin output does not affect the rate of NADH: ferricyanide-reductase reaction (since flavin is not a participant in the transfer of electrons from NADH to ferricyanide; flavin is needed for NADH: ubiquinone reductase activity, but not for NADH: ferricyanide reductase [60]). NADH: ferricyanide reductase reaction occurs mainly on the damaged complex I or a separate peripheral fragment [59].

The degree of decrease in the NADH oxidase activity with the loss of flavin was not so pronounced in comparison with a very strong increase in the intensity of flavin fluorescence. This is due to the fact that fluorescence quantum yield of flavin in water is many times greater than in MC, and the ubiquinone reductase activity of the enzyme, as is known [51–53, 61], is proportional to the content of flavin. Interestingly, the photo-destruction of mitochondrial flavins occurs faster than the photo-inactivation of NADH oxidase [60, 79].

NADH dehydrogenase activity of MC has an optimum at pH 6–8. At acidic (pH <5.5) and alkaline values (pH > 9), it drops sharply [59]. Acid or alkaline pH significantly enhances the release of flavins from MC (Table 6.5). This can be explained by the denaturation of proteins. Denaturation of complex I is accompanied by the loss of flavin.

When SDS detergent in a low concentration (less than 0.005%) was added, the MC membranes were a bit damaged, but not destroyed [59]). As a result, flavin fluorescence increases dramatically. SDS is capable to solubilize the peripheral fragment of complex I into a solution [59], as well as to interact with free flavin [82]. As can be seen from Table 6.5, the 0.001% SDS induces the discharge of flavin from the active center.

Detergents used in everyday life (“Fairy”, “Komet”, “Pemolux”, etc.) significantly increase the dumping of flavins (FMN and riboflavin) from MCs (Table 6.5). “Fairy” acts most quickly and strongly. Concentration of only 0.001% causes a total enlightenment of the sample, visible even with the naked eye. The intensity of the flavin fluorescence in this case increases sharply at the first minute of incubation. Sodium bicarbonate (soda) was the slowest and least active agent.

The data obtained indicate that the release of flavin (spontaneous loss, temperature-inducing loss, detergent-induced loss or others) from complex I occurs very easily and leads to decrease in NADH-oxidase activity. It is possible that many genetic and acquired defects of complex I, leading to diseases, are mediated through its conformational liability, leading to the loss of flavin. Effective detergents that remain in trace amounts on dishes are dangerous to humans.

6.4.1 Conclusion

Incubation of the MC suspension for 1–4 h leads to spontaneous release of flavins from MCs to the aqueous phase. When MCs are damaged by heating, traces of detergents or detergents, acidic or alkaline pH, the release of flavins is greatly enhanced. This is indicated by the increase in flavin fluorescence and its depolarization. The loss of flavins leads to the decrease in NADH: ubiquinone oxidoreductase activity, but not NADH: ferricyanide reductase. The results were published in [83].

6.5 Stabilization of Enzyme by Adenosine Phosphates

In recent years, senile manifestations of neuromuscular degenerative diseases are increasingly associated with impaired functioning of the NADH dehydrogenase complex I [84, 85] and, as a result, increased oxidative stress of cells. One of the factors causing these disorders is the loss of connection between flavin and the enzyme.

Flavin is located in the 50-kDa subunit of NADH dehydrogenase, in its active center (NADH sits nearby). Flavin is easily lost under various influences (in the form of FMN or riboflavin). This can occur spontaneously over time, at increasing temperature, under the influence of trace amounts of detergents, etc. [86, 87]. As a result, the electrons from NADH do not go further along the electron transport chain, but directly reduce oxygen to the superoxide anion. Overproduction of superoxide, which can lead to lipid peroxidation, to damage of proteins and DNA, causes the inevitable aging of both – the MC itself and other cellular organelles. During this process, cell death by apoptosis or necrosis may be triggered, depending on the extent of the MC damage.

To prevent the loss of connection between NADH dehydrogenase and flavin, one can use NAD, which stabilizes the enzyme [86, 87]. Such action of NAD is caused, firstly, by inhibition of the dehydrogenation reaction, as a product of reaction. Secondly, NAD mechanically stabilizes the active center that prevents the falling out of flavin. Thirdly, NAD can chelate iron clusters by own negative charges.

It was interesting to test some natural analogs of NAD like stabilizators of flavin in NADH dehydrogenase. For this, a number of substances with a chemical formula similar to NAD were selected: ATP, ADP, AMP, adenine, adenosine, nicotinic acid and nicotinamide.

In order to accelerate the spontaneous release of flavin from isolated MCs, they were incubated at 37 °C for 2 h. In order for better penetrating of NAD, NADH, adenine and adenosine inside the MC and for faster flavin exit, slightly hypotensive incubation medium containing 100 mM sucrose and 10 mM Tris-phosphate (pH 7.0) was applied. In such environment, membranes become more permeable. In addition, MCs in a number of experiments were taken after freezing, permeabilizing membranes.

To determine the enzymatic activity of NADH dehydrogenase, the photometric method with pNTV, which is reduced to formazan, was used [88]. This method is convenient in that the recovery rate of pNTV is ten times higher than the rate of electron transfer through the electron transport chain to oxygen, so it is not necessary to use any respiratory chain blockers.

The MC fraction was isolated from rat liver in a medium containing 20 mM Tris-HCl and 250 mM sucrose (pH 7.7) [89]. Mature heavy MCs were precipitated and then suspended in the same medium, packaged on ice in small test tubes, and used for experiments (often after freezing). In some cases, MCs were isolated in a medium containing 200 mM sucrose, 100 mM mannitol, 10 mM HEPES (pH 7.5); and double washing was performed with double centrifugation at 5000 g.

The oxidation of NADH by tetrazolium in MCs was detected by formation of formazan deposited inside the matrix (the suspension turned pink color). 100 μM NADH, 100 μM pNTV and the test substance (ADP, adenosine, etc.) in excess (up to 300 μM) were added to MCs. The kinetics of the increase of optical density during the formation of formazan was measured for 15 min at 540 nm. Protein concentration was determined by the UV method [90]. All samples were protein aligned using a buffer (100 mM sucrose and 10 mM Tris-phosphate, pH 7.0) to a concentration of 0.3 mg/ml.

Flavin fluorescence spectra were recorded on SLM-4800 in 1-cm or 0.4-cm cuvettes at 20 ° C (excitation 450 nm). When measuring the degree of polarization (P), the intensity was low; therefore, mirror cuvettes were used [87].

To detect the localization of the emerged flavin after 2-h incubation of MCs at 37 ° C, the suspension was passed through 0.2-μm millipore filters and then the fluorescence of the obtained filtrate was measured (MCs were deposited on the filter). In addition to flavins and flavoproteins, the embryonic PMCs also pass through 0.2-micron pores, but there are few of them and they contain few flavins [91, 92] and therefore practically do not affect the measurements. When MCs did not interfere with the measurement of flavin fluorescence, they were not filtered.

The process of flavin exit from the active center of NADH dehydrogenase can be accelerated by trace amounts of detergents or by prolonged heating. This is consistent with the well-known fact of complete or partial loss of flavin during preparative work during isolation and purification of NADH dehydrogenase [93–97]. Moreover, the loss of flavin is accompanied by a decrease in NADH: ubiquinone reductase activity, but not NADH: ferricyanide reductase, because flavin is not involved in the second reaction [87, 89].

In experiments on stabilization of flavin in the enzyme, the MC incubation was performed for 2 h at 37 °C. After that, flavin fluorescence was measured at 20 °C ($\lambda ex = 450$ nm, $\lambda em = 525$ nm). In addition, tryptophan fluorescence ($\lambda ex = 286$ nm, $\lambda em = 340$ nm) was measured to determine the release of proteins from MCs, as well as 7AAMD fluorescence ($\lambda ex = 530$ nm, $\lambda em = 620$ nm) to determine the release of DNA (7AAMD is built into the untwisted sections of DNA [87]). It turned out that during the release of flavin, there is no significant exit of proteins and DNA from MCs: the fluorescence intensity of proteins and 7AAMD in the medium after 2-h incubation of MCs at 37 °C and subsequent removal of MCs remained negligible. This means that the membranes of MCs remained mostly intact.

Moreover, the intensity of flavin fluorescence in control samples during incubation increased by 2–3 times. Some contribution to the total flavin fluorescence of the MC suspension was made not only by flavin belonging to NADH dehydrogenase, but also by FAD of other dehydrogenases. The change in the intensity of flavin fluorescence is apparently caused mainly by the release of FMN and riboflavin.

Fluorescence polarization degree was measured also. The polarization value was used to determine whether the flavin is bound or free [87]. The bound flavin has a high degree of polarization (due to the short lifetime and low rotational mobility inside the protein). As flavin enters the solution, the polarization degree decreases, since the rotation speed sharply increases in the aqueous phase and also increases the lifetime (due to the absence of quenching by iron-sulfur clusters). In the experiments, the polarization degree of flavin fluorescence after incubation at 37 °C after 2 h decreased by 1.5–2 times.

In order to verify exactly where the released flavin appeared – in the matrix or in an external solution – the fluorescence intensity and polarization degree were also measured in the filtrates obtained by filtering of MC suspensions (before and after 2 h of incubation) through 0.2 micron millipore filters. After incubation of MCs (in comparison with MCs before incubation), the intensity in the filtrate increased significantly – up to 4.3 times. Polarization degree in the filtrate decreased very strongly, to 0.06, which corresponds to the free rotation of FMN or riboflavin in aqueous phase [87]. The data obtained clearly indicate that almost all flavins enter in a free form into external solution.

Changes in flavin fluorescence are not related to the oxidation/reduction processes in flavins, since 2-h incubation of MCs in most experiments was carried out in the absence of NADH, succinate and other reducing agents or oxidizing agents. In only one case — when using NADH during incubation — two processes took place simultaneously: restoration of flavin to a non-fluorescent form and stabilization of flavin in the enzyme. Since the reduced flavin has loosening orbital and is not stable (it immediately gives electrons to the respiratory chain, to ubiquinone), it quickly transforms back into its oxidized fluorescent form. Therefore, flavin red-ox transitions are unlikely to make a significant contribution to changes of flavin fluorescence.

A number of substances with a chemical formula partially similar to NAD were tested: analogues of the adenyl part of the coenzyme – adenine, adenosine, ATP, ADP and AMP, as well as NADH analogues, which include the nicotinic part of the molecule – nicotinic acid, nicotinamide and NAD.

When ATP, ADP, or AMP was added at a concentration of 300 μM to suspension of MCs, flavin fluorescence decreased by 44%, 49%, and 41%, respectively, compared to the control (Fig. 6.7, Table 6.6).

In control experiments, when 300 μM AMP was added to 0.1 μM FMN dissolved in water (without MCs), the flavin fluorescence intensity did not change, i.e. there was no complexation of FMN with the tested substance. It should also be noted that AMP acted on the MC flavin fluorescence even in a 10-fold lower concentration (30 μM), which indicates a rather high constant for the binding of this molecule to NADH dehydrogenase.

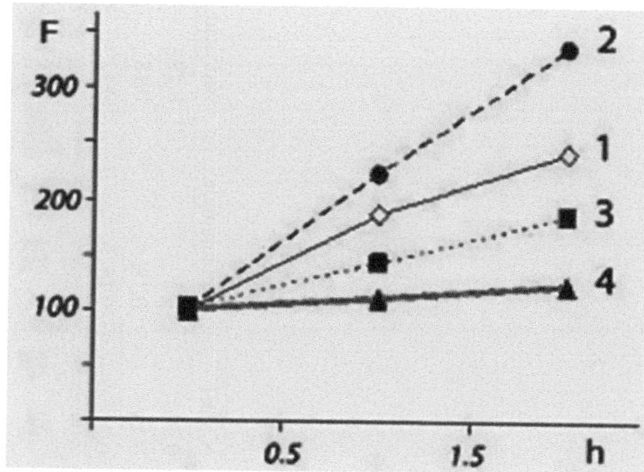

Fig. 6.7 The effect of NADH analogues on the intensity of flavin fluorescence during the during 1–2-h incubation of suspension of MCs at 37 °C: 1 – control, 2 – nicotinamide, 3 – NAD, 4 – AMP. The fluorescence intensity was measured at 20 °C

Table 6.6 The effect of NADH and its analogues (300 μM) on the flavin fluorescence intensity (%) during 1–2 h incubation of mitochondrial suspension at 37 °C

H	Control	Adenine	Adenosine	AMP	ADP	ATP	NAD	NADH	Nicotin amid	Nic. acid
0	100	100	100	94	99	102	101	101	100	101
1	186	203	179	130	110	132	142	160	223	219
2	241	307	256	142	122	133	185	199	336	322

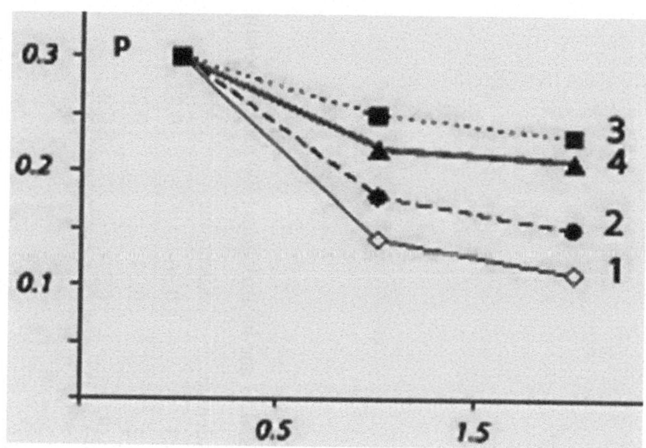

Fig. 6.8 The influence of NADH analogues on polarization degree of flavin fluorescence during incubation of mitochondria at 37 °C: 1 – control, 2 – nicotinamide, 3 – NAD, 4 – AMP, ADP and ATP. Measurements were done at 20 °C

The final fluorescence polarization degree of flavins upon tow-hours incubation of ATP, ADP, or AMP with MCs was 2, 1.9, and 1.8 times higher than the control (without added substances), respectively (Fig. 6.8, Table 6.7). This suggests that, in the presence of these substances, flavin molecules mainly remained bound to the enzyme.

Table 6.7 The effect of NADH and its analogues (300 μM) on polarization degree of flavin fluorescence

h	control	Adenine	Adenosine	AMP	ADP	ATP	NAD	NADH	Nicotinamide	Nic. acid
0	0.3	0.3	0.36	0.32	0.29	0.29	0.31	0.3	0.3	0.29
1	0.14	0.2	0.24	0.21	0.22	0.23	0.27	0.23	0.18	0.18
2	0.11	0.18	0.2	0.2	0.21	0.22	0.25	0.21	0.15	0.12

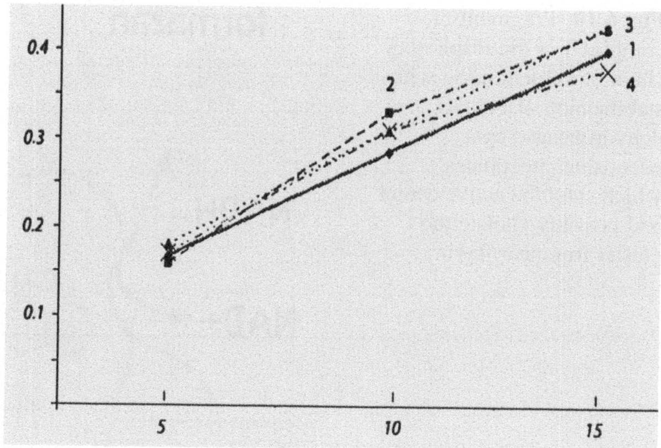

Fig. 6.9 The effect of adenosine phosphates on the reduction of pNTV to formazan. 1 – Control, 2 – AMP, 3 – ADP, 4 – ATP. Temperature 20 °C

Table 6.8 The oxidation rate of NADH, measured using pNTV by the accumulation of formazan in a suspension of mitochondria in the presence of 300 μM of test substances

Substance	Optical density D		
	5 min	10 min	15 min
Control	0.17	0.29	0.4
Adenine	0.16	0.27	0.31
Adenosine	0.16	0.27	0.31
AMP	0.16	0.33	0.43
ADP	0.18	0.31	0.43
ATP	0.17	0.31	0.38
NAD	0.12	0.15	0.19
Nicotin acid	0.13	0.24	0.34
Nicotin amide	0.16	0.24	0.34

It was also determined how these substances affect the enzymatic oxidation reaction of NADH. Figure 6.9 shows the kinetics of the NADH-tetrazolium reaction for 15 min. It can be seen that there are practically no strong deviations from the course of the reaction (compared with the control) in the case of ATP. Moreover, ADP and AMP even slightly (approximately 10–20%) accelerated the reaction. It can be concluded that these substances (ATP, ADP and AMP) do not inhibit the enzymatic activity (Table 6.8). In the used concentrations, they do not compete with NADH for the active center.

When adding adenine or adenosine at a concentration of 300 μM to the MC suspension, differences in the intensity of flavin fluorescence and polarization degree (Tables 6.6 and 6.7) were not very pronounced compared with the control. And these substances did not affect the enzymatic activity of NADH dehydrogenase (Table 6.8).

NAD and NADH inhibit the release of flavin molecules, but not as well as ATP, ADP, or AMP (Table 6.6). In high concentration, NAD successfully competes with NADH for the active center. At a concentration of 300 μM, it strongly suppresses the reaction (Table 6.8), and at a concentration of 2 mM it almost completely inhibits the formation of formazan.

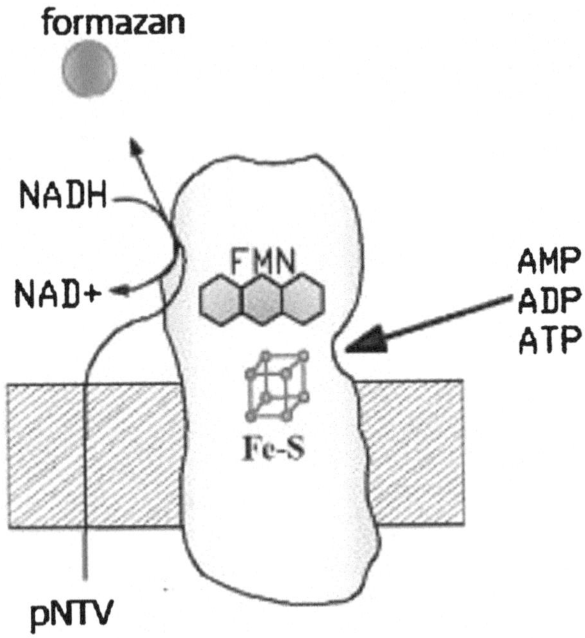

Fig. 6.10 Fragment of complex I of the respiratory chain of mitochondria with stabilization of NADH dehydrogenase by adenosine phosphates, which enter the active center and possibly chelate the cluster iron near flavin

A number of researchers [98–101] claim that there are two binding sites in NADH dehydrogenase (Fig. 6.10). In the first, NADH is oxidized, and in the second, NAD is reduced. It can be assumed that at the presence of NADH, the ATP, ADP and AMP molecules can bind to the second center, stabilizing it and preventing the release of flavin molecule from the protein in solution (the first center is occupied by the NADH or NAD molecule). Despite its small size (compared to NAD), adenyl nucleotides at a concentration of 300 μM do not compete for the binding site with NADH and do not inhibit the NADH: tetrazolium reaction (Table 6.8). In the case of AMP and ADP, this reaction is even accelerated by 30%, which may indicate allosteric regulation of the enzyme or some chelation of iron clusters by them. In the presence of ATP, the rate does not differ from the control. Adenosine phosphates either interact with the second center, or they have their own binding center on others enzymes.

Judging by the changes in flavin fluorescence, the effect of NAD was on 23% less than the control, which was on 21% less than that of ATP, ADP and AMP.

The highest degree of polarization of flavin fluorescence is observed in the presence of NAD, it is more than in the case of ATP (at concentrations of 300 μM).

When measuring the fluorescence intensity and polarization degree of flavins in the presence of NADH, it must be taken into account that, at the used concentration, NADH molecules in water could form dimers, which, excited at 450 nm, fluoresce in the same region (at 525 nm) as the FMN molecules (data not shown). At a concentration of 300 μM, NADH dimers contribute about 50% to the total intensity. Some researchers previously mistook the fluorescence of NADH dimers for flavin fluorescence. Excluding the contribution of NADH dimers, flavin fluorescence in the presence of NADH was 45% higher compared to control sample. ATP, ADP and AMP had the same difference – about of 45%.

When nicotinamide or nicotinic acid was added at a concentration of 300 μM to MCs, an increase in flavin fluorescence by 39% and 33% was observed (Figs. 6.6 and 6.7). Moreover, polarization degree differs little from the control. It can be concluded that these substances do not protect NADH dehydrogenase and other flavin-contained proteins from loss of flavins. They also do not affect the oxidation reaction of NADH (Table 6.8).

Summing up, we can conclude that adenosine phosphates (ATP, ADP and AMP) are able to effectively inhibit the release of FMN from NADH dehydrogenase and no too worse than NAD and NADH. They are hydrophilic substances, penetrating well into cells and MCs. As well known, cells usually maintain a constant ratio between ATP and ADP, and the content of AMP can vary markedly. The introduction of an excess of adenosine phosphates during treatment will not cause disturbances in the intracellular balance, but can temporarily activate adenosine phosphate-dependent enzymatic reactions. It is no accident that AMP and ATP are used in sports medicine [102], as well as for the treatment of muscular dystrophy, chronic coronary insufficiency, and myocardial dystrophy.

6.5.1 Conclusion

It can be assumed that mitochondrial encephalomyopathies (for example, Parkinson's disease) are most often caused by damage of NADH dehydrogenase that leads to loss of flavin from the enzyme. Isolated MCs easily lose flavin from the active center. The flavin release process is blocked by the substrate – NADH and the reaction product – NAD. Best of all, along with NAD and NADH, adenine derivatives such as ATP, ADP and AMP interfere with the release of flavin into the solution. The results were published in [103]. The data obtained can be used in the development of synthetic drugs similar to adenosine phosphates for the treatment and prevention of mitochondrial diseases.

6.6 Dehydrogenase Activity of Mitochondrial Smear, Detected by Decrease of Tryptophan Fluorescence by Formazan

In a living cell, there are many different membrane dehydrogenases that play a crucial role in metabolism. To judge the functioning of dehydrogenases in normal and pathological conditions, you need to be able to determine their activity. The activity of dehydrogenases during oxidation of substrates is usually measured colorimetrically by the restoration of artificial electron acceptors [104]. Moreover, the use of the method is limited by the sensitivity of a colorimeter or spectrophotometer. It is problematic to work on such devices with micro- and nanomolar concentrations. In many cases, it is necessary to use a minimum amount of sample with very low enzyme content, i.e. there is a need for a multiple increase in sensitivity.

Of all the instrumental methods of analysis, fluorescence spectroscopy is most sensitive [105], which is a thousand times more sensitive than photometry, NMR, ESR, etc. Since electron acceptors such as ferricyanide, DCPIP, tetrazolium salts, and many others are not capable of fluorescence, then it would seem that a fluorescent approach cannot be applied with them. However, as shown by the example of MCs, dehydrogenase activities can be determined by reducing in the intensity of protein fluorescence by formazan formed from pNTV [117]. This approach allows repeatedly increase the sensitivity. In addition, sensitivity is further enhanced when using a smear – by attaching the MCs to the center of the plate placed in the cuvette.

The method of colorimetric measurement of NADH: pNTV-reductase activity by the formation of formazan in the smear of MCs was considered in detail in the monograph [104]. The only drawback of the colorimetric method is not very high sensitivity. When the amount of MCs is small, the rate of formation of formazan is low, and its optical density is too low.

It was necessary to develop a highly sensitive method for measuring of NADH: pNTV-reductase activity in the MC smear so that it could be used to compare activities under different incubation conditions or different MCs in normal and pathological conditions. For enzymatic restoration of artificial acceptors by NADH, only three main subunits of complex I are sufficient [106, 107].

MCs were isolated from rat liver according to [108, 109], based on standard methods [110, 111], at 14000 g. The proto-mitochondrial (PMC) fraction was isolated from the MC suspension using millipore filters [112]. Optical microscopy of MCs stained with Janus green or formazan was performed. Protein concentration was determined by the UV method [113]. Kinetics of NADH: pNTV reductase reaction in the MC suspension was recorded on Specord UV-Vis in a special compartment for turbid samples. The reaction was carried out in thermostat with continuous stirring, taking small samples during the reaction to measure the optical density. Saturating concentrations of NADH and pNTV were used so that the initial rate depended only on protein concentration. The concentrations of NADH and pNTV in the experiments were 100 and 10 μM, respectively.

In the fluorimetric measurement of NADH: pNTV-reductase reaction, the MC smear was used, which was prepared according to [104] with modifications. A microdroplet (2–3 μl) of the MC suspension was placed on a defatted quartz plate and, using a covers-lip, smeared with a uniform thin layer. The smear was dried for several minutes, after which the plate was immersed for 1 min in isopropanol for better fixation. Then the sample was washed with water and dried again. Such a smear, when stored in the refrigerator, retains the NADH: pNTV reductase activity for many days.

The sample (MC smear on the plate) was placed for 1–10 min in an incubation medium containing 100 μM NADH, 10 μM pNTV, 250 mM sucrose and 20 mM Tris-HCl (pH 7.4) at 20 °C. The enzymatic reaction was stopped by transferring the sample to a medium (250 mM sucrose and 20 mM Tris-HCl, pH 7.4) without a substrate and an acceptor. This method of stopping the reaction eliminates the effects of screening and reabsorption (100 μM NADH and 10 μM pNTV are capable of strongly screening and reabsorbing light in the tryptophan excitation region at 280 nm and at emission band 340 nm). The baseline fluorescence level of the smear was determined in the absence of pNTV and NADH.

A decrease in protein fluorescence of smear during the NADH: pNTV reductase reaction was recorded on a Perkin-Elmer or SLM. Excitation monochromator slits – 2 nm, emission – 4 nm. In this case, a quartz 1 cm cuvette with medium (2–3 ml) was fixed in the instrument holder, and a quartz plate with a smear was placed diagonally in the cuvette.

6.6.1 NADH: pNTV Reductase Activity in the MC Suspension

The added NADH can be rapidly oxidized only in permeabilized MCs [114] (the "external" membrane NADH dehydrogenase in mammals is not normal [109, 115]). Therefore, the colorimetric determination of NADH: pNTV reductase activity was carried out in frozen/thawed MCs. Kinetics of NADH: pNTV-reductase reaction, detecting photometrically by accumulation of formazan in suspension of MCs, is shown in Fig. 6.11.

The oxidation rate of NADH decreased with the addition of an excess of NAD which suppressed the formation of formazan (Table 6.9). NAD competes with NADH for active centers and, being a reaction product, inhibits it.

As is known, tetrazolium dyes can be reduced by different reducing agents. However, if, in addition to NADH, other substrates are not added, then tetrazolium is reduced to formazan only through complex I (regardless of the presence or absence of rotenone), and other enzymes do not participate. The specificity of the reaction is confirmed by the fact that the process is strongly suppressed by the

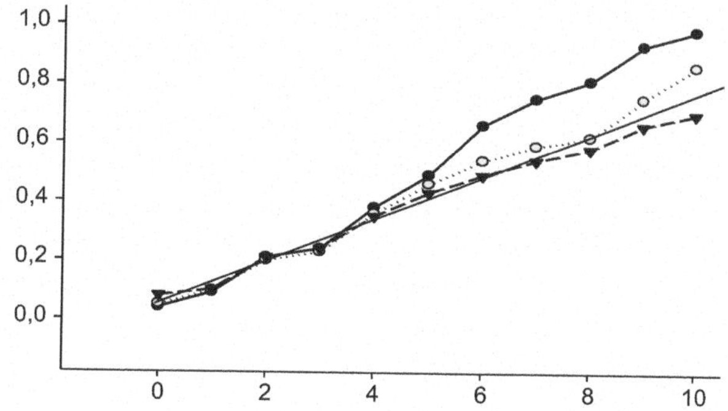

Fig. 6.11 NADH: pNTV reductase reaction at 540 nm (accumulation of formazan) in a suspension of MCs (1) and PMCs <0.45 μm (2) and PMCs <0.22 μm (3). Protein concentration was 0.1 mg/ml

Table 6.9 Dependence of the rate of recovery of pNTV to formazan in a suspension of mitochondria under different conditions

MCs + additives:	pNTV recovery rate,%
NADH, 100 μM	100 ± 5
NAD 1 mM and NADH 100 μM	20 ± 5
At 60 °C, then 100 μM NADH	85 ± 5
pH 3.0, then pH 7.4 and 100 μM NADH	5 ± 5
"Old" MCs and 100 μM NADH	50 ± 5
Succinate 5 mM	10 ± 2
Succinate 5 mM and malonate 5 mM	5 ± 2
Without substrates	0 ± 2

Note: Protein concentration was 0.1 mg/ml. Incubation time was 10 min. Colorimetry was done by accumulation of formazan

reaction product – an excess of NAD. In principle, tetrazolium can be reduced by other complexes, but this requires a special substrate. If only NADH is added, then the rate of the NADH: pNTV reductase reaction in complex I is so high that the restoration of tetrazolium by other complexes of the respiratory chain can be neglected.

Tetrazolium can also be reduced with superoxide. In our conditions, there is only one possible source of superoxide – NADH, which in complex I is able to reduce oxygen to superoxide. Since the oxygen concentration in the solution is high (~ 250 μM), and the solution in the cuvette is in contact with air, in the presence of 100 μM NADH, the reaction through the formation of superoxide will not limit the reaction rate and will not be able to mask the desired activity.

If succinate is added instead of NADH, then there will be succinate: pNTV-reductase reaction, which is rather slow compared to NADH: pNTV-reductase (Table 6.9). This reaction is specifically inhibited by malonate, but not NAD.

These reactions are not inhibited by rotenone. This means that they take place bypassing ubiquinone [109, 116]. According to [104], water-insoluble formazan formed from water-soluble tetrazolium is deposited exclusively in MCs. However, formazan appears not only inside the MC, but is able to crystallize outside of the MC, in the form of large aggregates. This is due to the fact that reactions with artificial acceptors occur not only inside the MC, but also on the surface of their fragments [109, 116]. Extra-mitochondrial granules of formazan were detected using light microscopy. This was also evidenced by the lack of reaching the plateau of kinetics of accumulation of formazan (Fig. 6.11). If formazan would be deposited only inside the MC, then gradually it would completely fill each MC and the reaction would slow down. But this does not happen even with a very long incubation, when the optical density of formazan at 540 nm reaches 1 (MCs are stained in bright pink color). Moreover, the same reaction does not reach a plateau in the case of PMC (Fig. 6.11) [112], the volume of which is several times smaller than the volume of MC. When large amounts of formazan are formed, it crystallizes not only inside MC organelles, but also outside.

Unfortunately, the colorimetric method is difficult to use at a protein concentration of less than 0.1 mg/ml. It is desirable to measure the enzymatic activity at low amounts MCs, PMCs and also at small accumulation formazan, when it is deposited mainly inside the MC, in matrix. It is important to expand the sensitivity threshold of the method. This cannot be done colorimetrically, but it is possible – fluorimetrically.

6.6.2 NADH: pNTV Reductase Activity, as Determined Fluorimetrically in a Smear of Mitochondria

NADH and pNTV at the used concentrations (100 and 10 μM, respectively) strongly shield and reabsorb light in the UV region, which makes quantitative measurements extremely difficult at the region of tryptophan excitation (280 nm) and emission (340 nm). For the correct fluorimetric measurement of NADH: pNTV reductase reaction, a mitochondrial smear was used.

The initial fluorescence intensity of the smear depends on many factors: the MC concentration, smear thickness, slit width of monochromators, etc. However, at the same conditions, this intensity and final intensity in various smears is within an error of 2–3%.

The fixation procedure does not lead to a loss of shunting activity of dehydrogenases [104], although the complete respiratory chain does not function. It was important to develop a highly sensitive method for measuring NADH: pNTV-reductase activity of the mitochondrial smear so that it could be used to compare the activities of complex I of MCs under different conditions (pH, temperature, etc.).

The application of MCs in the form of a smear on a quartz plate is accompanied by permeabilization of the membranes, as a result of which NADH from the incubation medium easily penetrates to NADH dehydrogenase. Therefore, hypotension or freezing/thawing of MCs are not required here. If such smear consisted of very small amount of MCs is placed in an incubation medium with NADH and pNTV, then formazan will begin to accumulate in matrix, which reduces the tryptophan fluorescence of MC proteins (Fig. 6.12).

Each point in Fig. 6.12 was obtained by incubating another individual sample. The indicated decrease in fluorescence is caused, first of all, by local micro-screening and micro-reabsorption of light [105] by formazan, which is deposited in matrix if each individual MC. Indeed, if we take the diameter of one MC equal to 10^{-4} cm, the extinction coefficient of formazan at 280 and 340 nm is about 10^4 M^{-1} cm^{-1}, and the local concentration of formazan inside a single MC is at least of 0.1 M, then the local optical density (in each individual MC) will be no lower than 0.1. This means that the tryptophan fluorescence intensity of each MC will be reduced by approximately 25% due to micro-screening and micro-reabsorption of light by formazan. If the concentration of formazan inside a single MC exceeds 1 M (this is crystalline formazan), then the local optical density will be greater than 1, i.e. the intensity of

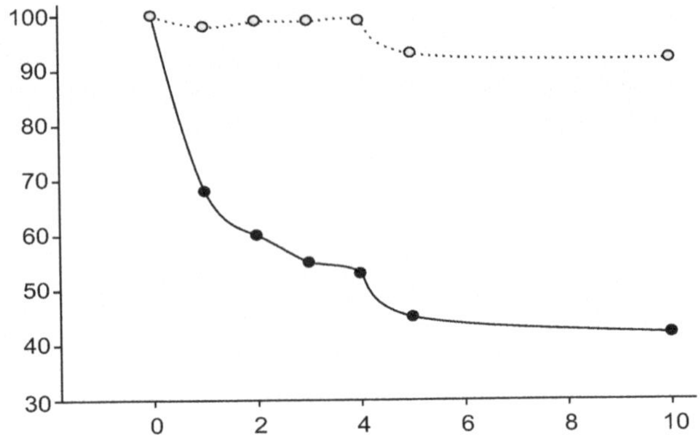

Fig. 6.12 Dependence of NADH: pNTV-reductase reaction on time (min), determined by the decrease in protein fluorescence of the MC smear during the 10-min accumulation of formazan. Protein concentration is only 0.01 mg/ml

Table 6.10 The influence of conditions on the value of NADH: pNTV-reductase reaction, measured by the reduction of tryptophan fluorescence of mitochondrial smear protein by formazan

Smear MCs + additives:	Fluorescence intensity, %
Control	100 ± 2
NADH, 100 μM	30 ± 2
NAD, 1 mM and NADH 100 μM	~ 100 ± 3
Incubation at 60 °C, then 100 μM NADH	85 ± 3
pH 3.0, then pH 7.4 and 100 μM NADH	~ 100 ± 3

Note: Control – fluorescence intensity before NADH: pNTV reductase reaction; experiments – after the reaction during 10 min (after reaching a plateau) under different conditions. Protein concentration was only 0.01 mg/ml

tryptophan fluorescence will decrease by no less than 90% compared to the initial intensity. With small amounts of formazan formed, the rule holds: the more tetrazolium molecules are reduced in the MC to formazan, the more tryptophan fluorescence decreases. In control experiments (incubation of smear without tetrazolium or without substrate), no significant changes in tryptophan fluorescence occurred during the first 4 min. A slight decrease in control after 4 min is apparently caused by some changes in the MC itself. The same decrease in the fourth minute is observed in the experiment. If we normalize the signals, take the experience/control ratio, then a small "failure" at 4 min becomes invisible.

NAD at a concentration of 1 mM strongly inhibited NADH: pNTV reductase activity (Table 6.10). This indicates the specificity of the reaction. In control experiments without a substrate, 1 mM NAD did not affect the fluorescence intensity.

Since the reaction takes place on the plate, the complete respiratory chain does not function: there is no electron transfer from complex I to other complexes. No breathing.

If succinate is taken instead of NADH, then tetrazolium (both in the presence of rotenone and in its absence) is also reduced to formazan, although much more slowly than with NADH. Succinate: pNTV reductase reaction can also be detected fluorimetrically.

Preliminary exposure of the smear for 10 min at pH 3 led, as expected, to a sharp suppression of enzymatic activity (Table 6.9), which was manifested by the absence of changes in fluorescence after incubation of the smear with NADH and pNTV (Table 6.10). This means that NADH: pNTV reductase is sensitive to pH inactivation.

Preliminary heating of the smear at 60 °C for 10 min or storage for 24 h at 20 °C resulted in only a 15% decrease in fluorescence by formazan. This means that the enzyme subunits responsible for NADH: pNTV reductase reaction are resistant to heat and long-term storage in the form of a smear. In the control, the effect of these factors on the smear (without an enzymatic reaction) reduced fluorescence by only 5–10%.

The maximum decrease in tryptophan fluorescence of mitochondrial smear with formazan after NADH: pNTV reductase reaction was no more than 60–70% (Fig. 6.12 and Table 6.10), and not 100%, as might be expected. This is due to the fact that a part of the proteins does not belong to MCs, but to impurity organelles (up to 30%), in which formazan does not form. The fluorescence of these organelles is about 30% of the total fluorescence and does not depend on intra-mitochondrial formazan.

In fluorimetric experiments, only 0.01 mg of protein in 1 ml was used. This is 10–100 times less than in the case of colorimetry method. The amount of formazan accumulated in the smear during the reaction was so insignificant that it was barely detected photometrically: the optical density at 540 nm in a 1-cm cell did not exceed 0.01.

So, the fluorimetric method allows you to measure trace amounts of dehydrogenases. The method allows accurate quantitative measurements. Moreover, the MC fluorescence intensity in the smear is several times greater than in the case of the MC suspension, since in the spectrofluorimeter the exciting light is focused by the lens into the center of the cuvette and is collected by the second lens also mainly from the center (through which the plate passes and where the smear is concentrated). With the correct position of the plate, the maximum fluorescence intensity is achieved, and the excitation light does not enter the recording channel and does not interfere with the fluorescence measurements.

6.6.3 Conclusion

A fluorimetric approach is proposed for detecting the activity of membrane dehydrogenases. For fluorimetric measurement of NADH: pNTV-reductase reaction, a fixed smear of MCs was used on a quartz plate, placed diagonally in a cuvette with a solution. Enzymatic activity was determined by the decrease of tryptophan fluorescence of mitochondrial smear proteins by formazan, which is formed during the restoration of pNTV. This decrease is due to local micro-screening and micro-reabsorption of fluorescence by formazan within each individual MC. The proposed fluorimetric method allows you to measure trace amounts of enzymes. The results were published in [117].

6.7 Rotenone-Insensitive Oxidation of NADH by Fragments of Respiratory Chain

Exogenous added NADH can be rapidly oxidized only in permeabilized MCs, for example, after they are frozen, damaged by any agents, or placed in a hypotonic medium [118]. The activation of NADH oxidation in MCs under the action of lipid peroxides is known [119].

Depending on the experimental conditions, the oxidation of NADH may be sensitive or not sensitive to rotenone, a specific inhibitor of NADH dehydrogenase of the respiratory chain [120, 121]. There was a point of view that the rotenone-insensitive (RI) oxidation pathway is carried out not through this NADH-dehydrogenase, but through a completely different enzyme located in the outer membrane [120, 122–126], or, as was suggested in [121], on the outer side of the inner membrane is the "outer path". On the other hand, it is noteworthy that most preparations of isolated NADH dehydrogenases are insensitive to rotenone [127–129]. A special check showed [130] that no "external" NADH dehydrogenase exists in intact MCs.

It was of interest to establish the cause of the occurrence of RI oxidation of NADH in the MC suspension. Ferricyanide was used as a model activator of RI oxidation [120, 121]. This substance is widely used as an electron acceptor to measure the activity of isolated NADH dehydrogenase [128, 129, 131, 132].

NADH dehydrogenase of respiratory chain catalyzes the rotenone-sensitive (RS) oxidation of NADH and the reduction of ubiquinone [133]. For this, iron-sulfur clusters, flavin, and phospholipids are needed [128, 129, 131]. There is a hypothesis that the enzyme functions as a dimer [134]. According to electron microscopy, complex I consist of two parts – membrane and peripheral [135]. When exposed to a chaotropic agent (0.5 M $NaClO_4$), the water-soluble part (~ 30% protein) is separated from the complex [128].

It was interesting to carry out a comparative study of the RS and RI oxidation of NADH in suspension of MCs (in a hypotonic environment).

The heart of a bull was taken within 2 h after the slaughter of the animal and placed in cubes of a frozen isolation medium containing 0.1 M Tris-HCl (pH 8.6 at 20 °C), 0.25 M sucrose and 1 mM succinate. MCs were isolated according to [139]. Since the minced buffer was acidified, the pH was quickly returned to 7.5 by the addition of Tris (pH 10.8). The mixture was suspended and homogenized for 1 min. The homogenate was adjusted to pH 7.8 and centrifuged for 15 min at 1500 g. The supernatant was diluted and centrifuged at 14,000 g for 20 min. The precipitate was suspended in a small amount of medium, carefully homogenized and diluted with 0.01 M Tris-HCl (pH 7.6), and centrifuged again at 14000 g for 20 min. The precipitate was suspended in several milliliters of medium. The resulting suspension of MCs ~ 8 mg/ml was aliquoted and frozen at −18 °C. Protein concentration was determined by the UV method [140]. Before experiments, all aliquots were thawed at 4 ° and placed on ice.

NADH dehydrogenase activity of MCs was measured in a hypotonic medium, containing 5 mM or 20 mM Tris-phosphate, pH 7.5. The MC suspension (0.025 or 0.25 mg protein per 1 ml) was placed in 0.5- or 1-cm quartz cuvettes in a special compartment of Specord UV Vis or M-40 for scattering samples. NADH oxidation was measured at 360 nm, where the contribution of ferricyanide was minimal. Reduction of ferricyanide to ferrocyanide was measured at 410 nm. The enzyme activities were determined by the initial rate: (a) during the first 30 or 60 s in the case of slow oxidation of NADH in respiratory chain, (b) during the first 10 s in the case of fast oxidation of NADH by ferricyanide. Excessive concentrations of NADH and ferricyanide (200 μM and 320 μM) were used, so that the initial rate depended only on the amount of protein.

Light scattering was measured in 1-cm cuvettes with "Specord UV Vis" at 360 nm by the difference in the optical densities of the suspension in a conventional compartment and in a special compartment for turbid samples.

A fraction containing detergent membrane fragments was obtained by filtering a suspension of MCs (in 2 ml of hypotonic medium containing 50 mM Tris, pH 7.5) at the presence of 0.01% SDS through a 0.2 μm FP 030/3 filter or Millex-GV.

The primary purification of detergent fragments of the enzyme complex from the filtrate was carried out on Sephadex G-50 columns. Elution was performed in 0.05 M Tris buffer, pH 7.5. The eluate was collected, photometrified on proteins at 280 nm, and fluorimetrified on flavins at 520 nm.

6.7.1 Oxidation of NADH in Suspension of Cardiac Mitochondria

When MCs were placed in a hypotonic medium (5 mM or 20 mM Tris-phosphate, pH 7.5) they oxidize the added NADH at a noticeable rate of ~ 0.1 μmol/min/mg protein. The observed oxidation was completely inhibited by rotenone or amital, as well as antimycin or azide (Table 6.11). This means that oxidation of NADH in used conditions occurs exclusively in the respiratory chain, and oxygen is consumed only at cytochrome oxidase level. There is no any "external path". The oxidation reaction is very specific: NAD**P**H (instead of NADH) does not oxidize (Table 6.11).

Rotenone is similar in chemical structure and hydrophobicity to ubiquinone. The sensitivity of the membrane NADH dehydrogenase to rotenone should be due to that rotenone is able to replace ubiquinone in hydrophobic binding sites, blocking electron transfer.

Table 6.11 The effect of various substances on the oxidation rate of added NADH (200 μM) in suspension of cardiac mitochondria in hypotonic medium (pH 7.5)

Additives	NADH dehydrogenase activity of respiratory chain, %	NADH: ferricyanide reductase activity, %
Rotenone 10 μM	0	~ 100
Amital 0.6 mM	0	100
Antimycin 10 μM	0	100
Azide 1 mM	0	~ 100
DTT 1.3 mM	85	14
NAD 5 mM	71	40
NADP 5 mM	85	96
EDTA 1 mM	100	100
EGTA 0,13 mM	~ 100	112
NADPH 200 μM, without NADH	~ 0	0
Glycerol 20%	~ 0	~ 100
SDS (0.01%)	~ 0	140
No additives	100	100

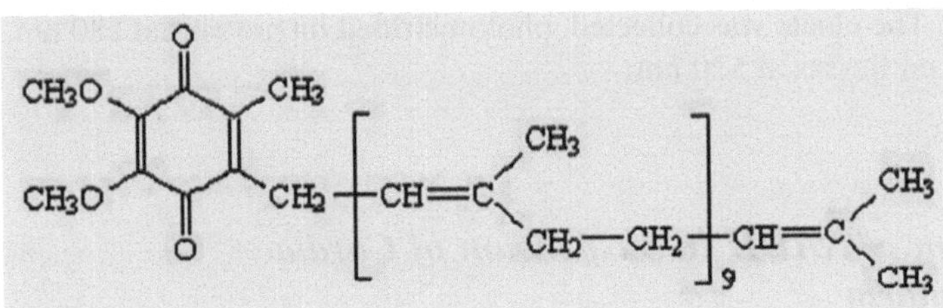

Ubiquinone

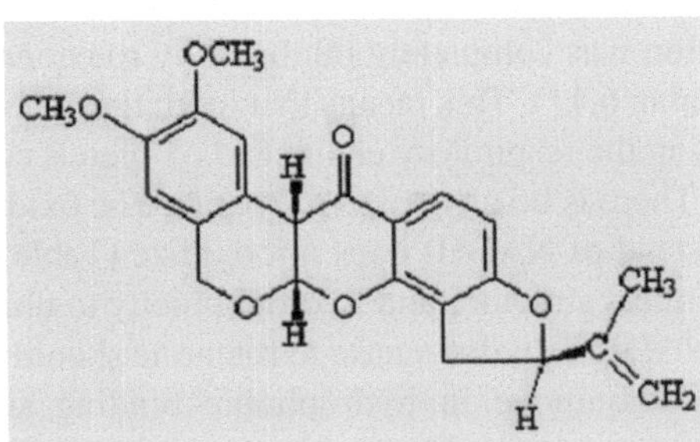

Rotenone

The NADH oxidation rate only slightly decreased with the addition of NAD or dithiothreitol (DTT) (Table 6.11). This means that in the membrane the active center of the enzyme and its S-S bonds are almost inaccessible. The decrease did not occur instantly, but a few seconds after the introduction of additives and practically did not change during incubation. 5 mM NAD inhibited the dehydrogenase activity in the

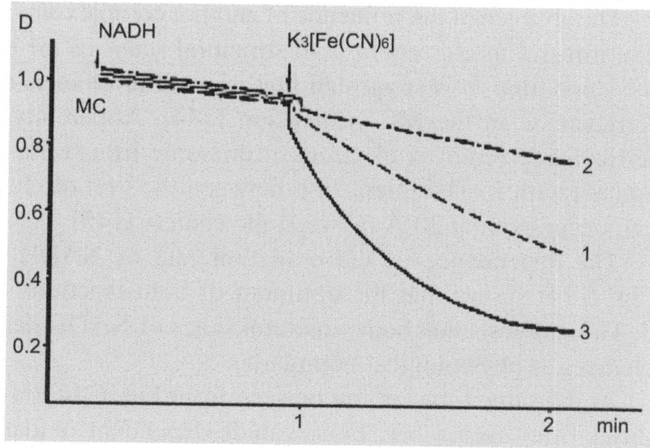

Fig. 6.13 Kinetics (at 360 nm) of oxidation of NADH in MCs (0.025 mg protein/ml) in a hypotonic medium at pH 7.5 in the presence of 5 mM NAD (1), 1.3 mM DTT (2) and without additions (3). Ferricyanide – 320 μM

presence of 0.2 mM NADH by only 29%. This not too strong competing effect of NAD is due to that the binding constant of NADH (10^6 M^{-1}) is an order of magnitude greater than NAD (10^5 M^{-1}) [141].

When ferricyanide was added to the suspension, the NADH oxidation rate increased sharply (Table 6.11, Fig. 6.13). The reaction is very specific: NAD**PH** does not oxidize with ferricyanide. The initial rate of enzymatic oxidation of NADH by ferricyanide was ~ 1 μmol/min/mg protein. In this case, ferricyanide was reduced to ferrocyanide (decrease of absorbance at 410 nm). Ferricyanide-induced oxidation of NADH is not sensitive to respiratory chain inhibitors (Table 6.11), i.e. the chain is not involved here. The lack of sensitivity to rotenone means that the oxidation of NADH is shunted to ferricyanide without participation of the ubiquinone-binding site (as it was observed with ferricyanide for an isolated enzyme [128, 131]).

The NADH: ferricyanide reductase reaction was strongly inhibited by the addition of DTT or NAD (Fig. 6.13), but not NAD**P** (Table 6.11). Inhibition of the reaction was observed immediately after addition and did not change during incubation. Since NAD is a reaction product, it greatly inhibits the reaction. DTT acts as a reducing agent for S-S bonds of the enzyme. Instant strong inhibition with the addition of DTT means that S-S bonds are placed near the surface. The labeling of some subunits of isolated NADH dehydrogenase with added DTT was shown in [142].

Significant differences in the degree of influence of DTT and NAD on the rates of RS and RI oxidation indicate that NADH dehydrogenase in the MC suspension exists in two structural states: "inaccessible" (for DTT and NAD) and "available". It cannot be ruled out that these two forms of the enzyme belong to two "populations" of mitochondrial particles.

Since the addition of EGTA – complexon on calcium – practically did not affect both types of oxidation (Table 6.11), this means that calcium ions released during hypotonic permeabilization of the membranes of cardiac MCs are not activators of NADH oxidation. This is fundamentally different from calcium-activated oxidation of endogenous NADH in hepatic MCs [143].

The absence of the influence of another complexon – **EDTA** (Table 6.11) – means that iron-sulfur clusters in both structural states of the enzyme are not available. At the same time, it is suggested that iron-sulfur clusters could be electron donors for ferricyanide in the MC suspension [144]. Apparently, in NADH dehydrogenase, ferricyanide removes electrons from some iron-sulfur cluster without direct close contact with it. The interaction between the iron of cluster and iron of ferricyanide can occur even at 20 Å between the centers [145].

The dependence of the oxidation rate of NADH on pH (curves 1 and 2 in Fig. 6.14) shows that the optimum of both reactions lies in the region from 6 to 8. This suggests that both structural states of NADH dehydrogenase are stable to pH changes at physiological boundaries.

At the same time, as can be seen from Fig. 6.14, the pH dependences noticeably differ from each other. Ferricyanide-dependent oxidation is more resistant to pH changes. It can be assumed that the reason for the sharper pH dependence of RS oxidation may be the ionization of respiratory chain proteins.

As well known, the functioning of dehydrogenases is suppressed in a viscous medium, for example, by the presence of glycerol [145, 146]. The addition of 20% glycerol (which increases the viscosity of the solution from 1 cps to just 1.8 cps) in the MC suspension led to almost complete inhibition of NADH oxidation in the respiratory chain, but had almost no effect on the rate of ferricyanide-induced oxidation of NADH (Table 6.11, Fig. 6.15). An increase in glycerol concentration to 33% (viscosity 2.8 cps) led to a twofold decrease in the reaction rate with ferricyanide.

Glycerol inhibition is not necessarily caused by an increase in the viscosity of the solution, since it may be due to the direct effect of glycerol on enzymes: the formation of hydrogen bonds with glycerol can increase the local viscosity on the surface of membranes. In any case, data on the action of glycerol suggests that the recovery of ferricyanide occurs with the participation of a different functional population of enzyme molecules than the restoration of mitochondrial ubiquinone.

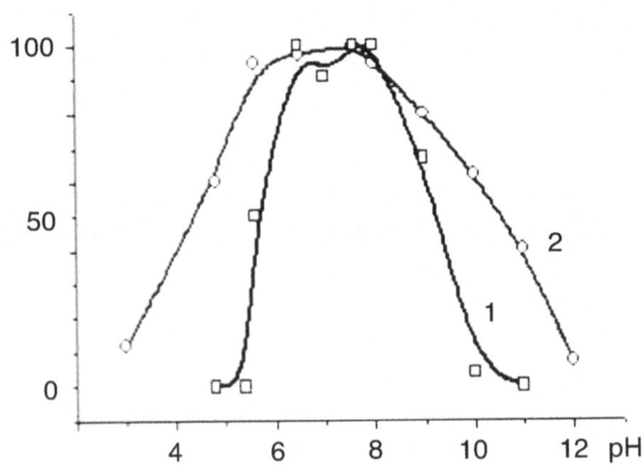

Fig. 6.14 Normalized dependences of the effect of pH on the oxidation rate of NADH in a suspension of mitochondria: by the respiratory chain (1) and a ferricyanide shunt in the same suspension (2)

Fig. 6.15 The effect of medium viscosity on oxidation of NADH by mitochondria. 1 – in the presence of ferricyanide, 2 – without ferricyanide

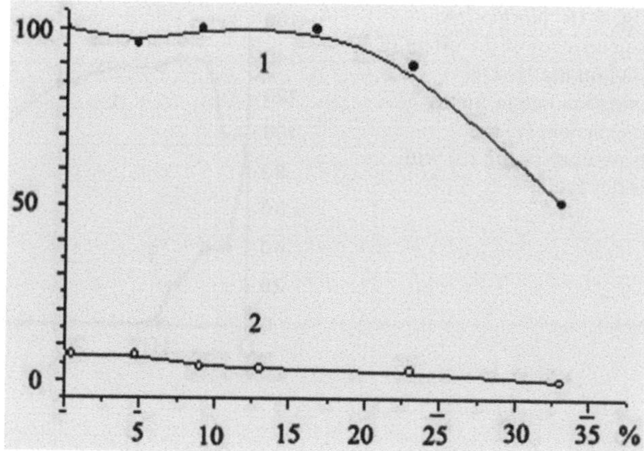

6.7.2 SDS Effect

The NADH oxidation reaction in the respiratory chain is completely inhibited even at low concentrations of SDS (Fig. 6.16), since the detergent disrupts the respiratory chain.

The addition of 0.005% SDS decreased the turbidity of the suspension by 30% (Fig. 6.17). This means that the average particle size was decreased. Between 0.005% and 0.016% SDS, the turbidity reached a plateau, i.e. here the particle size did not change. At 0.016%, a sharp enlightenment of the sample begins, associated with the complete destruction of MCs.

The dependence of the oxidation of NADH by ferricyanide on the concentration of SDS is complicated. As can be seen from Fig. 6.16, the detergent at a concentration of 0.01% activates the ferricyanide-dependent oxidation reaction of NADH by 40%. Probably, this effect is associated with the release (by detergent) of individual fragments of the enzyme complex or (and) of small fragments of membranes and, as a result, an increase in the steric availability of the enzyme for ferricyanide takes place. An increase in the concentration of SDS to ~ 0.02% almost does not affect the reaction rate, and an increase to 0.05% and higher inhibits it.

6.7.3 Obtaining Fractions of Detergent Membrane Fragments

The NADH: ferricyanide-active fraction containing detergent membrane fragments was prepared as follows. 80 μl of the MC suspension was diluted in 2 ml of hypotonic detergent buffer (50 mM Tris, pH 7.5 and 0.01% SDS), incubated for 2–3 min and filtered through a 0.2 micron FP 030/3 or Millex-GV filter (note: potassium phosphate buffer cannot be used here, since it forms strongly light

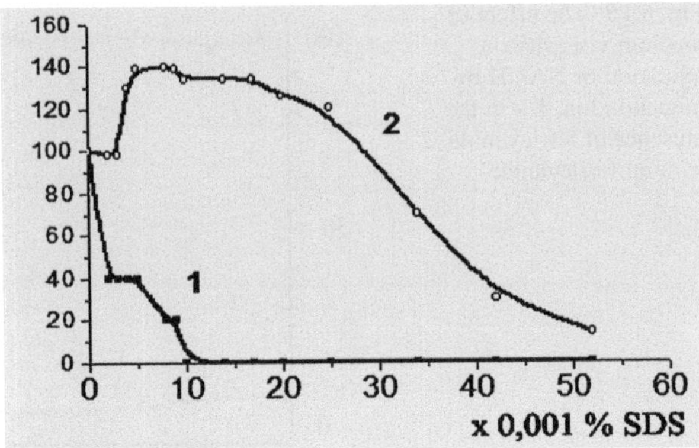

Fig. 6.16 Normalized dependences of the effect of SDS on the NADH oxidation rate in the MC suspension: (1) the respiratory chain, (2) with ferricyanide

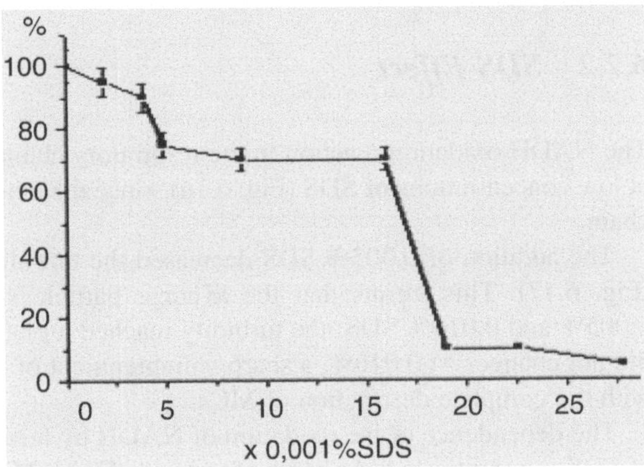

Fig. 6.17 Dependence of light scattering of a mitochondrial suspension (at 360 nm) on the concentration of SDS. Concentration of MCs – 0.3 mg in 1 ml

scattering micelles with SDS). At the indicated concentration, the detergent does not destroy the MCs, but only damages them. Since MCs have the 0.7–1.2-μm size [144], they linger on the filter. A transparent, slightly opalescent solution (filtrate) was passed through the filter. No RS-type oxidation was observed in the obtained detergent filtrate, but there was a large NADH: ferricyanide reductase activity, which reached 50% of the initial suspension activity. It is likely that SDS, replacing hydrophobic binding sites, promotes the separation of the peripheral fragment of the NADH-dehydrogenase complex or (and) individual membrane fragments from the inner membrane. When using some other filters (from other materials), the activity was significantly less or completely disappeared, for example, with a Sterifil GS filter.

In the absence of detergent, the same procedure gives a filtrate with three times less NADH: ferricyanide reductase activity and lower protein content. At the same time, a noticeable RS-type of NADH oxidation was observed, sensitive to antimycin and azide. This means that the obtained particles (PMCs) are smaller than 0.2 microns in size and have an active respiratory chain.

Table 6.12 Some properties of a detergent filtrate

Parameter:	%
Content of FMN	70
Content of riboflavin	15
Content of FAD	15
NADH: Ferricyanide reductase activity	100
The same activity with 5 μM FMN	100
The same activity with 650 mM KCl	38
The same activity after 24 h at 20° C	70

Some properties of the fragments contained in the detergent filtrate are shown in Table 6.12. The obtained fraction is water-soluble, active for NADH and artificial electron acceptors, stable in activity over time, resistant to freezing-thawing and long-term storage at −18 °C. Fragments are not only water soluble, but also alcohol soluble. When placed in 90% ethanol, the fragments remained in solution, did not aggregate, and even retained some enzymatic activity upon returning to the aqueous phase.

Incubation of detergent filtrate with added FMN did not affect the rate of ferricyanide-dependent oxidation of NADH, i.e. FMN is hardly important for this reaction. Ferricyanide-induced enzymatic oxidation of NADH occurs without the participation of flavin [147].

Using fluorescence and absorption spectroscopy it was established [148] that the filtrate contains the peripheral flavoprotein part of NADH dehydrogenase, as well as some quantity of cytochrome C and other proteins. No succinate dehydrogenase activity was detected.

6.7.4 Conclusion

A spectrophotometric method was used to carry out a comparative study of two types of NADH oxidation in a suspension of MCs from bovine heart in a hypotonic medium: RS and RI. Oxidation of added NADH in suspension occurs according to the RS type – through NADH dehydrogenase of the respiratory chain; it is completely inhibited by rotenone or amital, as well as antimycin and azide. Ferricyanide-activated oxidation of NADH is not sensitive to these inhibitors due to bypass electron transfer bypassing ubiquinone. Such RI oxidation, carried out by membrane fragments of NADH dehydrogenase, is strongly suppressed by NAD and DTT. RS-type oxidation is a bit suppressed by NAD and DTT. In the MC suspension, there are two structural populations of NADH dehydrogenase: “inaccessible” and “available”. For RS oxidation, the participation of the entire NADH dehydrogenase complex is necessary. Fragments are enough for the RI type. Low concentrations of SDS block the oxidation of NADH in the respiratory chain, but increase the rate with ferricyanide. This is caused by the release of membrane fragments. The results were published in [149].

6.8 Cytochrome C Shunts the NADH Oxidation

Flavin of NADH dehydrogenase is unlikely to be involved in fast shunt reactions. It was shown [150–152] that flavin and ubiquinone do not participate in the transfer of electrons to DCPIP and other artificial acceptors, i.e. artificial shunting of the electron beam occurs.

Shunting can occur also without artificial acceptors. One of the candidates for participation in natural shunting of electron transfer is cytochrome C.

Figure 6.18 shows that the addition of 60 μM cytochrome C increases the NADH oxidation rate in suspension of damaged (by freezing or hypotension) MCs more than double compared to the control. The addition of cytochrome C to MCs, the respiratory chain of which is blocked by rotenone, leads to 80% activation of the oxidation rate of NADH relative to the control level. This indicates the transfer of reducing equivalents bypassing the respiratory chain.

Measurement of the degree of reduction of cytochrome C by its absorption in the region of 530 nm showed that the addition of NADH to the MC suspension causes the reduction of cytochrome C, which after some time is replaced by its oxidation to the initial level (Fig. 6.19).

As can be seen from Fig. 6.19, the processes of reduction and oxidation of cytochrome C are separated in time (similar to the "lag period" in the reactions with DCPIP). This means that at first there is a rapid recovery of cytochrome C, and then it slowly diffuses to cytochrome oxidase, which oxidizes it.

Previously, some researchers suggested that there is a special enzyme diaphorase, having the NADH: cytochrome-C reductase activity. However, it can be argued that this reaction is simply bypassing electrons from NADH dehydrogenase of the respiratory chain to cytochrome C. The subsequent reaction is the oxidation of reduced cytochrome C by cytochrome oxidase.

Batophenanthroline (a blocker of iron-sulfur enzymes) and antimycin A prolong the lag phase that occurs after the NADH: cytochrome-C-reductase reaction (similar

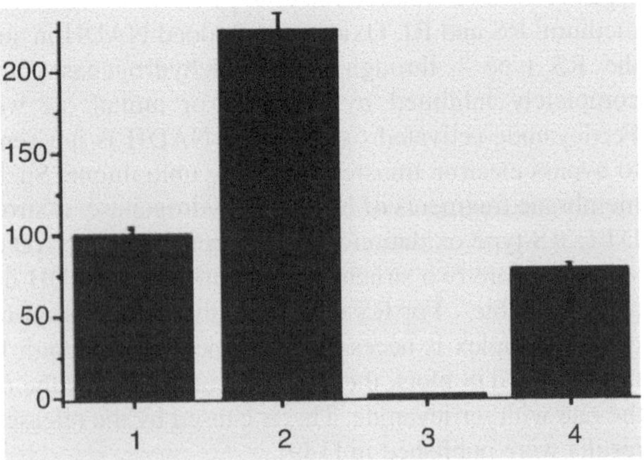

Fig. 6.18 The effect of cytochrome C on the NADH oxidase reaction in suspension of MCs. (1) NADH, (2) NADH + cytochrome C, (3) NADH + rotenone, (4) NADH + cytochrome C + rotenone. It was measured by the decrease in optical density of NADH

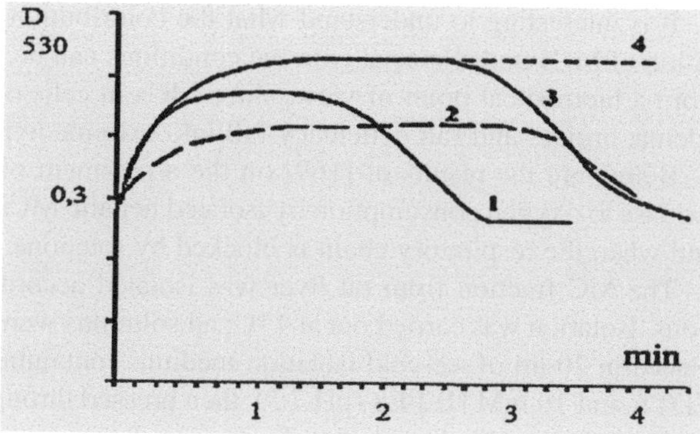

Fig. 6.19 The effect of inhibitors on the degree of reduction of cytochrome C in the process of NADH: cytochrome-C-reductase reaction. 1 – Control, 2 – in the presence of antimycin, 3 – in the presence of bathophenanthroline, 4 – in the presence of KCN

to the case with DCPIP). Cyanide completely inhibits the second reaction, which indicates the participation of cytochrome oxidase.

Thus, cytochrome C is able to shunt electrons from NADH dehydrogenase to cytochrome oxidase bypassing other components of the respiratory chain.

The obtained results confirm the work [153] on shunting electrons from NADH to cytochrome C and cytochrome oxidase, but do not confirm that some "special" NADH dehydrogenase of the outer membrane is involved. In fact, it is a common NADH dehydrogenase of a damaged inner membrane.

At the end of the twentieth century, it became known that MCs can activate cell apoptosis via releasing some signaling molecules. These "signaling molecules" turned out to be cytochrome C [154].

6.9 NADH Oxidase Activity of Mitochondria in Hypotension at Block of Respiratory Chain

During the oxidation of NADH in the respiratory chain under normal physiological conditions, oxygen consumption is carried out almost exclusively on cytochrome oxidase [155]. However, from 0.1% to 1% oxygen, depending on specific conditions, can still be consumed not on cytochrome oxidase, but directly on NADH oxidase [156, 157]. In this case, oxygen on NADH oxidase, taking an electron, is reduced to dangerous superoxide O_2^- [156–158]. When electrons are transferred back from other complexes of the chain to NADH dehydrogenase, up to 5% of oxygen uptake can be consumed directly on it [159]. When the respiratory chain is blocked by antimycin, production of superoxide is strongly increased [160].

Superoxide is very toxic and plays a significant role in pathologies and aging. It is believed that electronic leaks mainly occur on the Q_{in} site. The addition of antimycin contributes to this process. Antimycin is thought to block cytochrome b hem, preventing it from dropping electrons to the semiquinone Q•, which reacts with oxygen to form $O2^-$. However, there is a concern that more serious electronic leaks may occur at the NADH level in the active center of NADH dehydrogenase.

It is interesting to understand what the contribution of NADH oxidase to respiration of MCs under special extreme conditions can be. This is especially important from a biomedical point of view, since MCs in cells of organs with inflammation, edema, bruises and salt deficiency fall into extreme hypotonic conditions.

Below are the results of [169] on the assessment of the contribution of NADH oxidase to oxygen consumption by isolated hepatic MCs under hypotonic conditions and when the respiratory chain is blocked by rotenone, antimycin, and azide.

The MC fraction from rat liver was isolated according to [161] with modifications. Isolation was carried out at 4 °C; all solutions were cooled on ice. Rat liver was placed in 20 ml of ice-cold isolation medium containing 250 mM sucrose, 0.5 mM EDTA and 10 mM HEPES (pH 7.5), then pressed through a press, 80 ml of the same medium was added and the liver was homogenized. The homogenate was centrifuged for 10 min at 1000 g to precipitate intact cells and red blood cells. The pellet was discarded, and the supernatant was centrifuged for 15 min at 3000 g (heavy fraction of MCs). The pellet was carefully homogenized in 5–8 ml of isolation medium. The obtained MCs were stored on ice and used for experiments for several hours. The concentration of MCs was found by the UV method [162].

The isotonic incubation medium contained 150 mM sucrose, 10 mM Tris, 5 mM NaH_2PO_4, 1 mM EGTA (pH 7.5). The hypotonic incubation medium contained 10 mM Tris, 5 mM NaH_2PO_4, 1 mM EGTA (pH 7.5). In experiments with complete hypotension, pure distilled water, without additives, was used. The hypertonic medium contained 300 mM sucrose, 10 mM Tris, 5 mM NaH_2PO_4, 1 mM EGTA (pH 7.5).

Measurement of oxygen consumption by suspension of MCs in a 1.5-ml hermetic cuvette with stirring (in a hypotonic, hypertonic, or isotonic medium with succinate or NADH added) was carried out using a Clarke electrode or a Hamilton sensor.

Concentrations of antimycin – up to 5 μM and azide – up to 5 mM were applied for MCs with 3 mM succinate in isotonic medium, achieving maximum respiratory suppression. These concentrations of respiratory poisons were further used to inhibit the oxygen uptake arising from oxidation of NADH (800 μM) in a suspension of MCs in hypotonic medium or water. The concentration of rotenone – up to 5 μM was selected to maximize blocking the oxidation of NADH directly in the presence of NADH (without succinate).

The kinetics of oxygen consumption in suspension of MCs was measured under different conditions (Figs. 6.20, 6.21, 6.22, 6.23, 6.24 and 6.25, Table 6.13). As a criterion for the effectiveness of respiratory poisons, the ratio of the rate of oxygen consumption when adding a substrate (succinate or NADH) and poison (azide, rotenone, antimycin) to the rate when adding only a substrate was used. This criterion was called residual respiration.

The addition of succinate to MCs leads [163] to oxidation of these substrate by succinate dehydrogenase, followed by electron transfer through the respiratory chain to cytochrome oxidase, where oxygen is reduced to water. The oxygen consumption observed in an isotonic medium with succinate is slightly accelerated in a hypotonic medium, but is suppressed in a hypertonic medium (Fig. 6.20, Table 6.13). In hypotension, the permeability of MC membranes for the substrate increases, and in hypertension it decreases. Respiratory rate is limited by the passive permeability

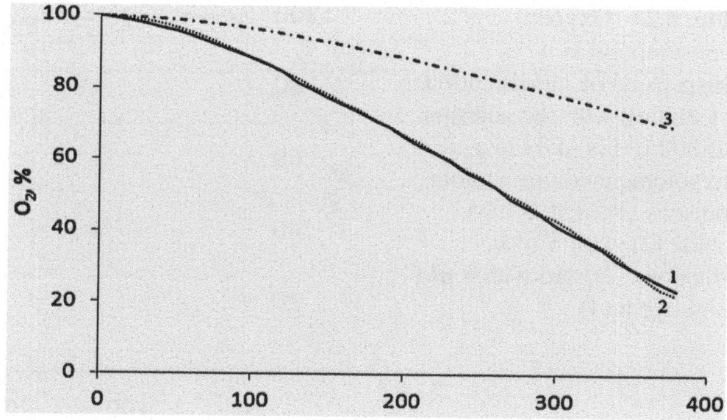

Fig. 6.20 Oxygen consumption (during 400 sec) in suspension of mitochondria (1 mg/ml) with the addition of 5 mM succinate in isotonic medium (1), in hypotonic medium (2) and in hypertonic medium (3)

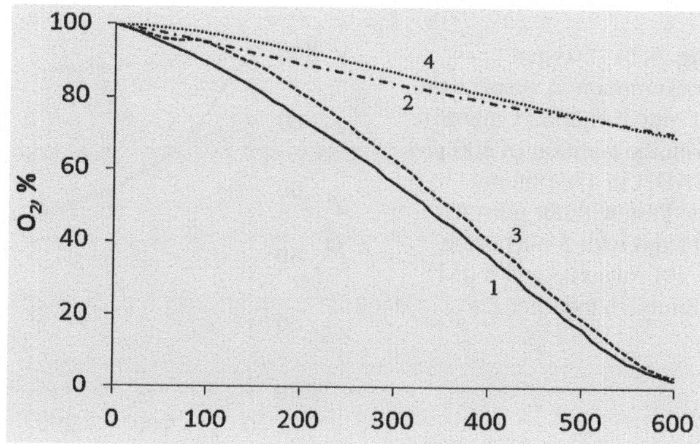

Fig. 6.21 Oxygen consumption in suspension of mitochondria (2 mg/ml) with the addition of 5 mM succinate in isotonic medium without poisons (1), with 5 mM azide (2), with 5 μM rotenone (3), with 5 μM antimycin (4)

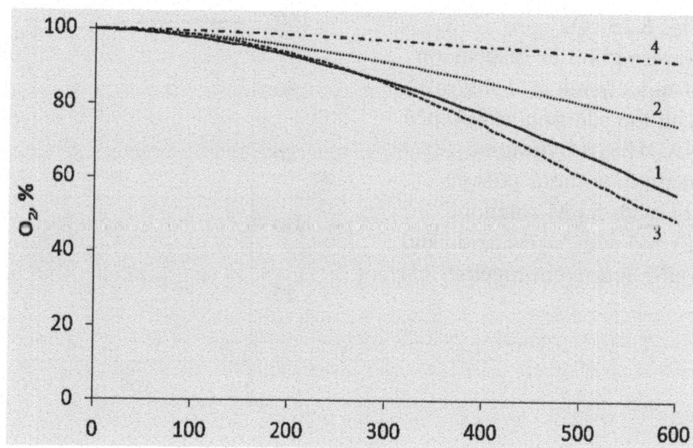

Fig. 6.22 Oxygen consumption in a suspension of mitochondria (1 mg/ml) with the addition of 5 mM succinate in a hypotonic medium without poisons (1), with 5 mM azide (2), with 5 μM rotenone (3), and with 5 μM antimycin (4)

of the membranes for succinate entering inside simply along the gradient, apparently without the participation of any special translocase carrier.

Succinate-dependent oxygen consumption is suppressed, as is known [155], by respiratory poisons, for example, antimycin and azide, the first of which blocks electron transfer at the level of cytochromes B and C, and the second – at the level of cytochrome oxidase.

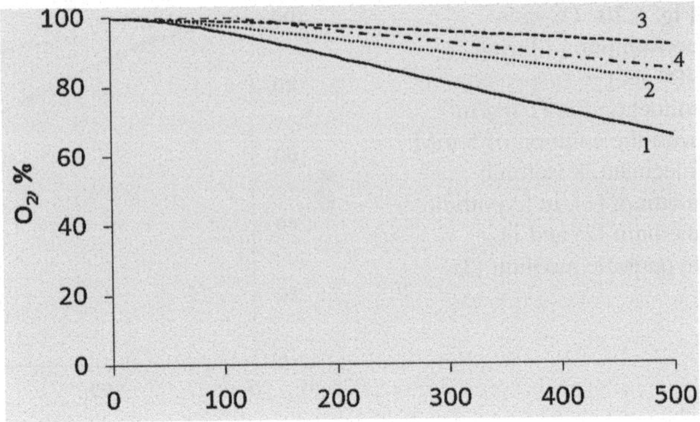

Fig. 6.23 Oxygen consumption in a suspension of mitochondria (1 mg/ml) with the addition of 800 μM NADH in a hypotonic medium without poisons (1), with 5 mM azide (2), with 5 μM rotenone (3), and with 5 μM antimycin (4)

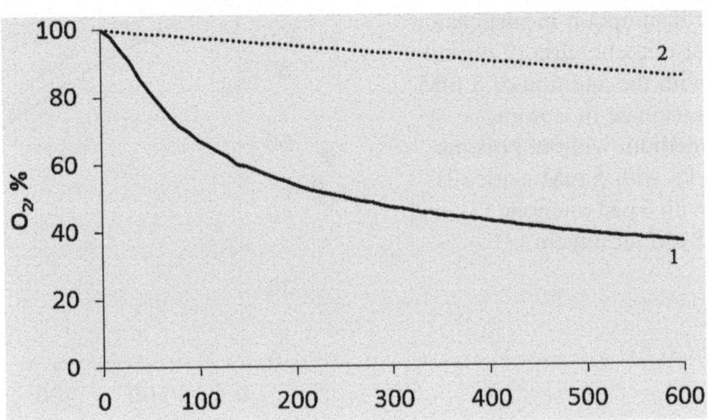

Fig. 6.24 Oxygen consumption in suspension of mitochondria (1 mg/ml) with the addition of 800 μM NADH in a hypotonic medium without poisons (1) and with 5 mM azide, 5 μM rotenone and 5 μM antimycin together (2)

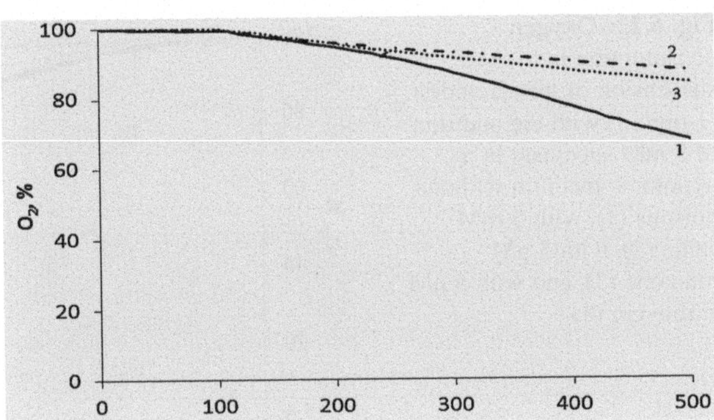

Fig. 6.25 Oxygen consumption in suspension of mitochondria (1 mg/ml) with the addition of 800 μM NADH in a hypotonic medium without poisons (1), with 5 μM rotenone (2) and with 5 mM azide and 5 μM antimycin together (3)

Antimycin has a slightly greater inhibitory effect on the succinate-dependent respiration of MCs (residual respiration was 20%) than azide (residual respiration was 30%) (Figs. 6.21 and 6.22). Moreover, in hypotension, antimycin inhibited respiration almost completely (Fig. 6.22). Rotenone practically did not affect succinate-dependent respiration, as expected.

Rotenone interacts with the ubiquinone binding site [155]. The well-known absence inhibition of respiration on succinate by rotenone means that (a) either

Table 6.13 The rate of oxygen consumption by suspension of mitochondria (1 mg/ml) in various conditions

		Mitochondria							
		NADH				Succinate			
O_2 µM /min/ mg			Azide	Rot	Antim		Azide	Rot	Antim
	Water	0.07	0.055	0.03	0.03	0.34	0.05	0.25	0.05
	Hypotension	0.085	0.045	0.025	0.035	0.11	0.05	0.11	0.02
	Isotension	0.01				0.09	0.03	0.11	0.025
	Hypertension	0.02				0.05			

electron transfer from succinate dehydrogenase downstream occurs without the participation of ubiquinone, (b) or rotenone is not a competitor of ubiquinone in NADH dehydrogenase, which is unlikely.

The addition of NADH to freshly isolated MCs in isotonic medium did not lead, as one would expect, to its oxidation (Table 6.13). The reason for this is the low permeability of the inner membrane for NADH [155]. In isotonia, the oxidation of NADH in the MC suspension took place only when membranes were damaged by freezing, etc.

In hypotonic medium, rapid oxidation of NADH by freshly isolated MCs is observed (Table 6.13), since permeability of the inner membrane for NADH appears.

Interestingly, in a hypertonic environment (residual respiration here was 24%), NADH oxidation also occurred. Apparently, in a hypertonic environment, the inner membrane is damaged by high ionic strength and therefore becomes permeable to NADH.

When NADH is used as a substrate in a hypotonic medium, rotenone significantly inhibits electron transfer from NADH dehydrogenase to ubiquinone [164], but 29% residual respiration remains.

In a hypotonic environment, the oxidation of NADH was accompanied by almost the same strong oxygen consumption as in the case of oxidation of succinate (Table 6.13). The speed of the process is limited by the passive permeability of the membranes for NADH, which enters inward along the gradient (without any participation of a special translocase transporter). It must be emphasized that MCs in hypotension are not native organelles, but they are only shadows. In them, the outer membrane is partially destroyed, and a significant part of the matrix has come out [165].

NADH dependent respiration is suppressed, as is known [164], by rotenone, which blocks the transfer of electrons from the iron-sulfur clusters of NADH dehydrogenase to ubiquinone. Rotenone strongly suppressed NADH-dependent oxygen consumption (Fig. 6.23). Once again, it is useful to recall that there is no "external" (in the outer membrane) NADH dehydrogenase [166].

It can be seen from Fig. 6.23 that in a hypotonic medium, antimycin or azide does not completely suppress the NADH dependent MC respiration. Antimycin suppresses more strongly (residual respiration 41%) than azide (residual respiration 53%).

It is noteworthy that in hypotension even the simultaneous combination of three poisons does not completely inhibit the oxygen uptake process (Fig. 6.24). As can be seen from Table 6.13, inhibition by poisons in moderate hypotension occurs more actively than in complete hypotension (in water).

It would seem that two poisons – antimycin and azide – should do the same thing as rotenone – to block the chain, completely eliminating the respiration. However, in hypotension, they suppress NADH-dependent respiration only partially, even when both poisons are used simultaneously (Fig. 6.25). This means that the sequence of electron transfer from NADH to oxygen here is completely different than from succinate. In other words, residual respiration not inhibited by antimycin and azide does not occur on cytochrome oxidase, but – directly on NADH oxidase itself. Its value reaches 44%. This automatically means the formation of a large amount of superoxide on NADH oxidase (Fig. 6.26).

So, the amount of superoxide formed on NADH dehydrogenase in MCs should increase very significantly under extreme conditions: in a hypotonic environment and when blocked by respiratory poisons.

The data on strong inhibition of succinate-dependent respiration by antimycin and azide, but weaker inhibition of NADH-dependent respiration by these poisons, suggest that the respiratory chain is not linear, but branched, i.e. consists of at least two parallel chains connecting at the level of cytochromes, and not at the level of ubiquinone. Two chains are less vulnerable than one.

It should be noted here that B.Chance's spectroscopic experiments [164, 167] on isolated and intracellular MCs, which formed the basis of the points of application of respiratory poisons, were not entirely correct. Thus, it was not taken into account that although rotenone practically does not fluoresce in aqueous solutions, but it can strongly fluoresce in the region of 440 nm upon binding to MCs—upon excitation at 340 nm—in the same place where NADH is excited. Another inhibitor, antimycin, alone does not absorb at 340 nm and does not fluoresce, however, upon binding to MCs and generally to protein-lipid complexes, it begins to absorb (at 320–370 nm) and strongly fluoresce (at 400–450 nm). Figure 6.27 shows the fluorescence excitation spectrum of the antimycin complex with albumin; an intense band is detected at 350 nm.

Another inhibitor used by B.Chance, cyanide itself does not absorb light and does not fluoresce, but it is able to enter into a chemical reaction with pyridine ring of NAD, forming an absorbing and fluorescent adduct, very similar in spectral properties to NADH [168].

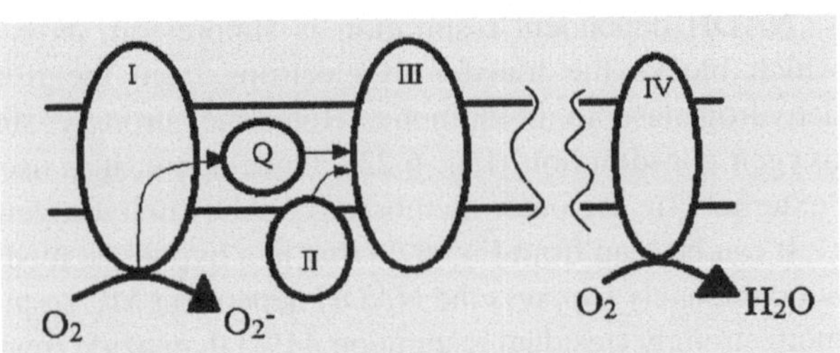

Fig. 6.26 Points of oxygen consumption in the native and damaged respiratory chain

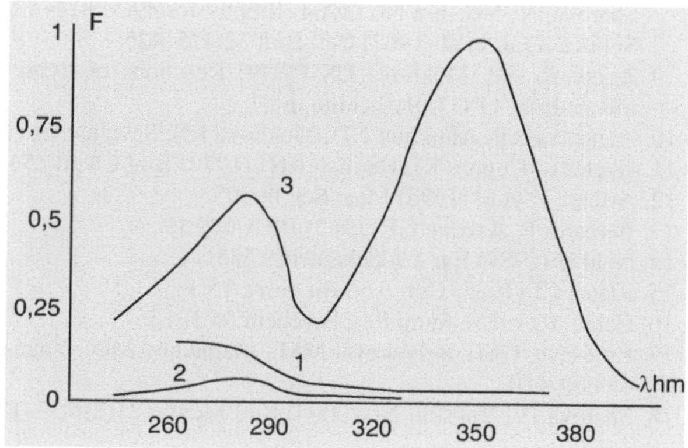

Fig. 6.27 The excitation spectrum of 3 μM antimycin (1), 0.2 mg/ml bovine albumin (2) and their complex (3). Emission was at 440 nm

In short, the introduction of cyanide, antimycin, or rotenone in the experiments of Chance [164, 167] and others could create the illusion of restoring NAD to NADH.

6.9.1 Conclusion

The question on the contribution of NADH oxidase to respiration of the MC suspension under extreme conditions (in a hypotonic environment and when blocked by respiratory poisons – rotenone, antimycin and azide) was considered. It was shown that significant oxygen consumption during oxidation of added NADH at these conditions occurs not only on cytochrome oxidase, but also on NADH oxidase. That is why NADH-dependent respiration is not completely suppressed by poisons. Respiratory activity insensitive to poisons is the result of electron shunting from NADH to oxygen directly on NADH dehydrogenase [169]. This residual respiration on NADH oxidase, not suppressed by the inhibitors, can under extreme conditions reach up to half of all oxygen consumption, giving superoxide.

References

References for Section 6.1

1. Vinogradov AD (1979) Reactions of living systems and the state of energy metabolism. ONTI, Pushchino, pp 98–125
2. Sharova IV, Vekshin NL (2001) Mitochondria in pathology. NCBI, Pushchino, pp 227–229
3. Mosolova IM, Gorskaya IA, Scholz KF, Kotelnikova AV (1975) Methods of modern biochemistry. Nauka, Mjscow, pp 45–47
4. Panov AV (2015) Practical mitochondriology. Nauka, Novosibirsk
5. Zelensky MI (1986) Polarographic determination of oxygen. Nauka, Leningrad
6. Vekshin NL (2002) Photonics of biopolymers. Springer, Berlin

7. Sharova IV, Vekshin NL (2004) Biophysics 49(5):814–821
8. Sottocasa GL et al (1967) Cell Biol 32:415–425
9. Agureeva AP, Mokhova EN (1979) Reactions of living systems and the state of energy metabolism. ONTI, Pushchino, p 27
10. Agureeva AP, Altukhov ND, Mokhova EN, Saveliev IA (1981) Biochemistry 46:1945
11. Hagler L, Coppes RI, Herman RHJ (1979) Biol Chem 254:6505
12. Arihara K et al (1995) Meat Sci 39:205
13. Bernardi P, Azzone GF (1982) BBA 679:19
14. Nohl H (1987) Eur J Biochem 169:585
15. Ragan CJ (1985) Curr Top Bioenerg 15:1
16. Hatefi Y (1985) Annu Rev Biochem 54:1015
17. Kolesova GM, Belyakova MM, Mamedov MD, Yaguzhinsky LS (2000) Biochemistry 65:680–684
18. Sharova IV, Vekshin NL (2004) Biol Membr 21(2):133–137

References for Section 6.2

19. Hatefi Y (1985) Annu Rev Biochem 54:1015–1025
20. Brandt U (1997) BBA 1318:79–91
21. Boekema EJ et al (1984) BBA 787:19–26
22. Dooijewaard G, Slater EC (1978) BBA 503:405–424
23. Walker JE, Skehel JM, Buchanan SK (1995) Methods Enzymol 260:14–34
24. Vekshin N, Sharova I, Sukharev V (1999) In: Ghisla S (ed) Flavins and flavoproteins. Elsevier, Berlin, pp 635–638
25. Lomtev AS, Sharova IV, Vekshin NL (2001) Dokl Akad Nauk 376(1):114–116
26. Dorovska-Taran V, Van Hoek A (1994) FEBS Lett 348:305–310
27. Vekshin NL (1999) Photonics of biopolymers. MGU, Moscow
28. Hink MA, Van Hoek A (1999) Langmuir 15:992–997
29. Sharova IV, Vekshin NL (2000) Mitochondria, cells and reactive oxygen species. ONTI, Pushchino
30. Panov AV (2015) Practical mitochondriology, Novosibirsk
31. Hatefi Y, Rieske JS (1967) Methods Enzymol 10:225–239
32. Valeur B, Weber G (1977) Photochem Photobiol 25:441–444
33. Sukharev VI, Vekshin NL (2000) Bioorg Chem 26:723–726

References for Section 6.3

34. Katsumata Y, Miyazi A et al (1970) J Biochem 68:423–435
35. Chernitsky EA (1972) Luminescent and structural lability of proteins in solution and cell. Nauka, Minsk
36. Sorokovoi VI, Klebanov GI, Vladimirov YA (1972) Mol Biol 6:705–711
37. Chernogryadskaya NA, Rozanov YM, Bogdanova MS, Borovikov YS (1978) Ultraviolet fluorescence of a cell. Nauka, Leningrad
38. Vekshin NL (1988) Photonics of biological structures. Academy of Sciences of the USSR, Pushchino
39. Chance B, Baltscheffsky H (1958) J Biol Chem 233:736–740
40. Scholz KF, Babadzhanova RG (1971) DoklAkad Sci 199:478–479

41. Mosolova IM, Gorskaya IA, Scholz KF, Kotelnikova AV (1975) Methods of modern biochemistry. Nauka, Moscow, pp 45–47
42. Vekshin NL (1988) Biol Sci 4:107–111
43. Lakovich J (1985) Fundamentals of fluorescence spectroscopy. Mir, Moscow
44. Vekshin NL (1989) Results of science and technology. Ser. "Radiation chemistry and photochemistry".7. VINITI, Moscow
45. Schellenberg KA (1965) J Biol Chem 240:1165–1169
46. Sankaran B et al (1996) Biochemistry 35:13501–13510
47. Vekshin N (1998) J Prikl Spectr 65(5):794–798
48. Leikin YN, Vinogradov AD (1974) Regulation of oxidation and conjugation. Nauka, Pushchino, pp 55–66
49. Agureev AP, Altukhov ND, Mokhova EN, Saveliev IA (1981) Biochemistry 46:1945–1956

References for Section 6.4

50. Hatefi Y (1985) Annu Rev Biochem 54:1015
51. Paech C, Friend A, Singer T (1982) Biochem J 203:477
52. Walker JE et al (1995) Methods Enzymol 260:14–34
53. Hirst J, Carroll J, Fearnley I, Shannon R, Walker J (2003) Biochim Biophys Acta 1604:135–150
54. Mitochondrial Disfunction BMF (1995) Oxidative damage in neurodegenerative diseases. Springer, Berlin
55. Cooper JM et al (1990) Biochem Soc Trans 18:517–519
56. Lenaz G et al (1997) Mol Cell Biochem 174:329–333
57. Wallace DC (1999) Science 283:1482–1488
58. Swerdlow RH et al (1998) Ann Neurol 44:873–881
59. Sharova IV, Vekshin NL (2004) Biophysics 49(5):814–821
60. Lomtev AS, Sharova IV, Vekshin NL (2001) Dokl Acad Nauk 376(1):114–116
61. Dooijewaard G, Slater E et al (1978) BBA 503:405–424
62. Almeida T et al (1999) Eur J Biochem 265:86–92
63. Krishnamoorthy G, Hinkle PCJ (1988) Biol Chem 263:17566–17575
64. Melov S, Coskun P et al (1999) PNAS 96:846–851
65. Glinn M, See C, Ervster L (1997) BBA 3118:264–254
66. Schapira AHV (1998) BBA 1364:261–270
67. Morgan-Hughes JA, Schapira AHV, Cooper JM, Clark JB (1988) J Bioenerg Biomembr 20:365–382
68. Tanaka M et al (1986) BBRC 40:88–93
69. Ichiki T et al (1988) Ann Neurol 23:287–294
70. Mizuno Y, Ohta S et al (1989) BBRC 163:1450–1455
71. Hattori N et al (1998) Genomics 49:52–58
72. Bentlage HACM et al (1995) BBA 1234:63–73
73. Pulkes T et al (1999) Ann Neurol 46:916–919
74. Esposti DM (1998) BBA 1364:222–235
75. Ushakiva A, Grivennikova VG, Ohnishi T, Vinogradov AD (1999) BBA 1409:143–153
76. Okun JG, Zickermann V, Zwicker K, Schagger H, Brandt U (2000) BBA 1459:77–87
77. Belyakovich AG (1990) Study of mitochondria and bacteria using pNTP tetrazolium salt. ONTI, Pushchino
78. Vekshin NL (1988) Biol Nauk 4:107–111
79. Vekshin NL (2006) Fluorescence spectroscopy of biopolymers. Photon, Pushchino
80. Lomtev AS, Bobrov AG, Vekshin NL (2004) Bioorg Chem 30(3):218–223

81. Abraham R, Kotlyar AB (1991) Biochemistry 56:2253–2260
82. Sukharev VI, Vekshin NL (2000) Bioorg Chem 26(10):723–727
83. Sokolova IB, Vekshin NL (2008) Biophysics 53(1):73–77

References for Section 6.5

84. Seo BB et al (2004) Hum Gene Ther 15(9):887–895
85. Varghese M et al (2009) J Neurol Sci 279(1–2):39–42
86. Sokolova IB, Vekshin NL (2008) Biophysics 53(1):73–77
87. Vekshin NL (2009) Fluorescence spectroscopy of biopolymers. Photon, Pushchino. 176 p
88. Belyakovich AG (1990) Study of mitochondria and bacteria using the tetrazolium salt of pNTV. ONTI, Pushchino
89. Sharova IV, Vekshin NL (2004) Biophysics 49(5):814–821
90. Vekshin NL (1988) Biol Sci 4:107–111
91. Zinina AN, Vekshin NL (2008) Biol Membr 25(6):480–487
92. Vekshin NL (2010) Biol Membr 27(5):424–429
93. Galante YM, Hatefi Y (1979) Arch Biochem Biophys 192:559–568
94. Dooijewaard G, Slater E et al (1978) BBA 503:405–424
95. Almeida T, Durante M, Melo AMP, Videira A (1999) Eur J Biochem 265:86–92
96. Cremona T, Kerney EB (1964) J Biol Chem 230(7):328–333
97. Okun JG, Zickermann V, Zwicker K, Schagger H, Brandt U (2000) Biochim Biophys Acta 1459:77–87
98. Grivennikova VG, Vinogradov AD (2003) Success Biol Chem 43:19–58
99. Vinogradov AD (1998) BBA 1364:169–185
100. Grivennikova VG, Vinogradov AD (2006) Biochim Biophys Acta 1757:553–561
101. Albracht SPJ, Mariette A et al (1997) BBA 1318:92–106
102. Bubnova TV (2009) Approved drugs and dietary supplements. Pen-Publ, Penza, 36 p
103. Frolova MS, Vekshin NL (2014) J Fluoresc 24:1061–1066

References for Section 6.6

104. Belyakovich AG (1990) Study of mitochondria and Bacteria using Tetrazolium salt pNTV. ONTI, Pushchino
105. Vekshin NL (2002) Photonics of biopolymers. Springer, Berlin
106. Dooijewaard G et al (1978) BBA 503:405–424
107. Walker JE, Skehel JM, Buchanan SK (1995) Methods Enzymol 260:14–34
108. Vekshin NL, Sharova IV (2004) Biol Membr 21(2):133–137
109. Sharova IV, Vekshin NL (2004) Biophysics 49(5):814–821
110. Panov AV (2015) Practical mitochondriology. Nauka, Novosibirsk
111. Mosolova IM, Gorskaya IA, Scholz KF, Kotelnikova AV (1975) Methods of modern biochemistry. Nauka, Moscow, pp 45–47
112. Shishmakov DA, Anisimov RL, Vekshin NL (2004) Biol Membr 21(5):374–380
113. Vekshin NL (1988) Biol Sci 4:107–111
114. Leninger A (1966) Mitochondria. Mir, Moscow. p. 230
115. Fraisse G, Rey E et al (1993) BBA 1143:190–197
116. Lomtev AS, Sharova IV, Vekshin NL (2001) Dokl Acad Sci 376(1):114–116
117. Sokolova IB, Vekshin NL (2006) Biol Membr 23(3):217–222

References for Section 6.7

118. Lehninger A (1966) Mitochondria. Mir, Moscow
119. Green RC, O'Brien PJ (1973) BBA 293:334
120. Sottocasa GL et al (1967) Cell Biol 32:415
121. Nohl H (1987) Eur J Biochem 169:585
122. Agureev AP, Mokhova EN (1979) Reactions of living systems and the state of energy metabolism. ONTI, Pushchino, p 27
123. Agureev AP et al (1981) Biochemistry 46:1945
124. Hagler L, Coppes RI, Herman RHJ (1979) Biol Chem 254:6505
125. Arihara K et al (1995) Meat Sci 39:205
126. Bernardi P, Azzone GF (1982) BBA 679:19
127. Ragan CJ (1985) Curr Top Bioenerg 15:1
128. Hatefi Y (1985) Annu Rev Biochem 54:1015
129. Walker JE, Skehel JM, Buchanan SK (1995) Methods Enzymol 260:14
130. Fraisse L, Rey E, Rigoulet M (1993) BBA 1143:190
131. Paech C, Friend A, Singer TP (1982) Biochem J 203:477
132. Sharova IV, Vekshin NL (2000) Mitochondria, cells and reactive oxygen species. ONTI, Pushchino, pp 163–166
133. Brandt U (1997) BBA 1318:79
134. Albracht SPJ, De Jong AMP (1997) BBA 1318:92
135. Boekema EJ et al (1984) BBA 787:19
136. Fry M, Green DEJ (1981) Biol Chem 256:1874
137. Weiss H et al (1991) Eur J Biochem 197:563
138. Finel M, Skehel JM, Albracht SPJ, Fearnley IM, Walker JE (1992) Biochemistry 31:11425
139. Panov AV (2015) Practical mitochondriology. Nauka, Novosobirsk
140. Vekshin NL (1988) Biol Sci 4:107
141. Abraham R, Kotlyar AB (1991) Biochemistry 56:2253–2260
142. Gondal JA, Anderson WMJ (1985) Biol Chem 260:12690
143. Vekshin NL (1998) Biochemistry 63:S. 1301
144. Belyakovich AG (1990) Study of mitochondria and bacteria using the tetrazolium salt pNTV. ONTI, Pushchino
145. Marcus RA, Sutin N (1985) BBA 811:265–322
146. Demchenko et al (1989) BBA 998:196
147. Lomtev AS, Sharova IV, Vekshin NL (2001) Rep Acad Sci 376:114–116
148. Vekshin NL, Sharova IV, Sukharev VI (1999) In: Ghisla S (ed) Flavins and flavoproteins. Elsevier, Berlin, p 635
149. Sharova IV, Vekshin NL (2004) Biophysics 49(5):814–821

References for Section 6.8

150. Lomtev AS, Sharova IV, Vekshin NL (2001) Dokl Acad Sci 376(1):114–116
151. Sharova IV, Vekshin NL (2004) Biol Membr 21(2):133–137
152. Sharova IV, Vekshin NL (2004) Biophysics 49(5):814–821
153. Bernardi P, Azzone GFJ (1981) Biol Chem 256:7187–7192
154. Vier J, Linsinger G, Hacker G (1999) BBRC 261:71–78

References for Section 6.9

155. Lehninger A (1974) Biochemistry. Mir, Moscow
156. Hansford RG, Hogue BA, Mildaziene VJ (1997) Bioenerg Biomembr 29(1):89–95
157. Murphy MP (2009) Biochem J 417(1):1–13
158. Turrens JF, Boveris A (1980) Biochem J 191(2):421–427
159. Muller FL (2008) et al. Biochem J 409(2):491–499
160. Boveris A, Cadenas E (1975) FEBS Lett 54(3):311–314
161. Frolova MS, Vekshin NLJ (2014) Fluores 24:1061–1066
162. Vekshin NL (2002) Photonics of biopolymers. Springer, Berlin
163. Skulachev VP, Bogachev AV, Kasparinsky FO (2010) Membrane bioenergetics. Moscow State University, Moscow
164. Chance B, Hollunger GJ (1963) Biol Chem 278(1):418–431
165. Kurdyukov DN, Vekshin NL (2016) Biophysics 61(4):736–743
166. Sharova IV, Vekshin NL (2004) Biophysics 49(5):814–821
167. Chance B, Baltscheffsky HJ (1958) Biol Chem 233:736–740
168. Yakovlev VA (ed) (1973) NADH and its analogs. Medicine, pp 3–150
169. Lvov A.M., Vekshin N.L. 2019, Biophysics., in press

Chapter 7
Proto-mitochondria

7.1 Some Properties of Proto-mitochondria

MCs of different sizes and with different properties are organelles at different stages of their life cycle. Let us give some examples confirming this.

On Neorospora crassa, it was shown that the growth of mitochondrial organelles occurs by adding new components [1, 2].

The oocytes of the early stages of oogenesis contain mainly small mitochondrial organelles with single cristae [3]. They are located in the perinuclear zone of the oocyte and have a spherical shape with a diameter of 0.2–0.5 microns [4].

Within 7 days of the development of chicken liver, the size of mitochondrial organelles increases fivefold [5].

The differentiated adult MCs lose their ability to grow and share [4].

In a living cell, as a rule, three populations are simultaneously present at the same time: young proto-MCs (PMCs) with a diameter substantially less than 1 micron, mature MCs – about 1 micron and old MCs – larger than 1 micron, moreover, the ratio of populations depends on the type of cells, age and a number of other parameters [6]. In a living cell, young small PMCs (0.2–0.4 μm) gradually grow into adult MCs (~ 1 μm) [6]. Mammalian PMCs were firstly isolated and studied in [7–9].

Since one of the most important functions of MCs is membrane transport of electrons and protons, it seems likely that chemical or structural differences in the membranes of organelles at different stages of development will affect the activity of the respiratory chain enzymes and other properties. Therefore, a comparative study of some of properties of PMCs obtained by filtering the total mitochondrial suspension (from rat liver or bovine heart) through 0.22 and 0.45 μ millipore filters was carried out.

MC fractions from bovine heart or rat liver were isolated using standard methods [10, 11] with modifications. Isolation was carried out at 4 °C in 0.1 M Tris-HCl (pH 7.4) containing 0.25 M sucrose. First, the homogenate (heart or liver) was

N. Vekshin, *Biophysics of Mitochondria*,
https://doi.org/10.1007/978-3-030-33853-4_7

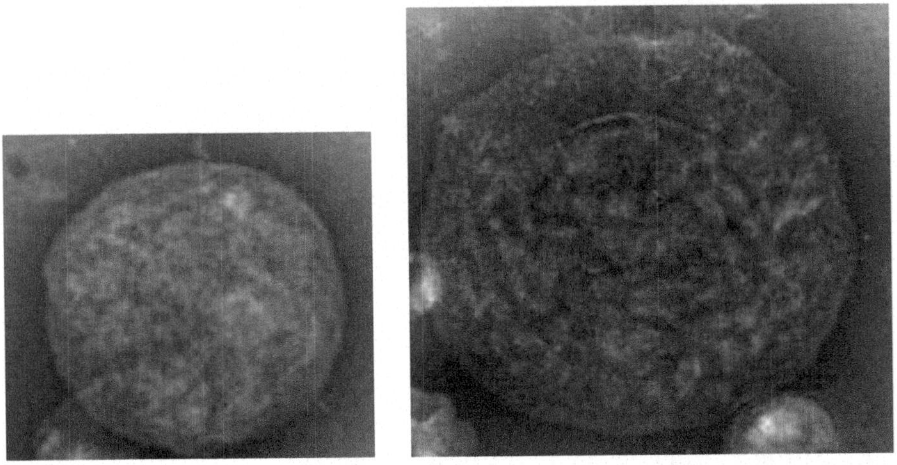

Fig. 7.1 Electron microscopy (negative contrast) of bovine hart proto-mitochondria with a diameter of 0.2 microns and 0.45 microns

centrifuged for 15 min at 1500 g. The supernatant was diluted with isolation medium and centrifuged at 10,000 g for 20 min. The precipitate was suspended in several ml of this medium. The resulting suspension was aliquoted. Aliquots from bovine heart were frozen at −18 °C. The suspension from rat liver was stored on ice. Protein concentration was determined by the UV method [12] and a standard biuret method.

Separate PMC fractions were obtained according to [7, 13] by filtering a diluted mitochondrial suspension (diluted in the isolation medium to a concentration of 3 mg/ml) through Millipore filters FlowPoreD "Sartorius" and FP030/3 "Schleicher & Schuell" with pore diameters of 0.45 and 0.22 microns. Particle size was controlled by correlation microscopy [8], as well as transmitted light microscopy – by staining with Janus green or formazan, which was formed upon enzymatic oxidation of succinate in MCs by pNTV [3]. The number of MCs was counted using optical microscope with an increase of ×640.

Electron microscopy showed that PMC with a diameter of 0.2 μm have only one membrane and do not have cristae, while PMC with a diameter of 0.45 μm have two membranes and developed cristae [17] (Fig. 7.1).

Fluorescence spectra of tryptophan residues of proteins, endogenous NADH (together with NADPH) and flavins were recorded on Perkin-Elmer MF44 using excitation at 280, 340 and 450 nm, respectively.

NADH oxidation was observed at three samples: a total MC suspension, PMCs up to 0.45 microns in size and PMCs up to 0.22 micron.

A small part of the proteins in the fractions belongs to ballast proteins released from MCs during the preparation of fractions, as well as non-mitochondrial organelles (microsomes, etc.). The specific activity calculated by the protein content in the fractions does not reflect the true specific activity of the enzyme. Therefore, the catalytic activity of PMCs was determined taking into account the content of ballast proteins, the concentration of which was measured in the supernatant obtained by centrifugation of PMCs. From the activity values, the specific activity of PMCs with diameter of 0–0.2 μm and PMCs 0.2–0.45 μm in diameter was determined in comparison with the total suspension of MCs.

The total cytochrome content was measured on "M-40" spectrophotometer (in a special compartment for scattering samples) by the ratio of optical density in the Soret band at 410 nm to the protein concentration. The relative DNA content was determined from the fluorescence intensity of ethidium bromide on "SLM-4800" in mirror micro-cuvettes [12] (excitation – 530 nm, emission – 600 nm).

NADH dehydrogenase activity was measured at 20 °C in isotonic (250 mM sucrose) or hypotonic (without sucrose) solution containing 20 mM Tris-HCl (pH 8.0). The reaction was carried out in 1-cm quartz or acrylic cuvettes in a thermostatically controlled compartment of "M-40". The oxidation of NADH in the presence of ferricyanide was detected photometrically at 360 nm. The initial optical density of NADH at 360 nm did not exceed 0.8. Activity was determined by the initial reaction rate: during the first 30–60 s in the case of oxidation of NADH in the respiratory chain and 10–20 s in the case of NADH: ferricyanide reductase reaction.

The rates of succinate oxidation and ADP phosphorylation were measured by oxygen uptake using a Clark electrode. The respiratory control (RC) was determined by the ratio of the oxygen consumption rate during ADP phosphorylation to ATP to the initial oxygen consumption rate during succinate oxidation (before ADP was added). The reaction mixture contained 5 mM succinate, 200 μM ADP, 2 mM phosphate.

The protein profile of organelles was obtained by electrophoresis under denaturing conditions in PAAG. The gel was applied with mitochondrial fractions (120 μg/ml) solubilized by SDS.

7.1.1 Fractionation of a Suspension of Mitochondria on Filters

The yield of proto-mitochondrial fractions during Millipore filtration of the total MC suspension depends on various conditions, including the concentration of MCs before fractionation, temperature, etc. However, under standard conditions, the obtained values of the enzymatic activities of PMCs in individual fractions in different series of experiments were very similar. Before applying to the filter, the initial suspension of MCs was diluted to a concentration of less than 3 mg/ml (otherwise the filter clogs and the yield of particles drops sharply). The distribution of total proteins in fractions is shown in Table 7.1.

The total cytochrome content was similar in all fractions. The DNA content was also not significantly different; the difference between the maximum and minimum fluorescence of ethidium bromide in fractions was only 8%. The data obtained indicate the similarity of all fractions (Table 7.2).

Table 7.1 The distribution of total proteins in fractions of mitochondria isolated from bovine heart

Fraction	Protein yield (%)	Particle size in a fraction (μm)	Percents of particles of a given size (%)
Mitochondria	100	>0.45	67
		0.2–0.45	21
		0–0.2	12
PMC 0–0.45 μm	58 ± 2	0.2–0.45	64
		0–0.2	36
PMC 0–0.2 μm	42 ± 0.6	0–0.2	100
Ballast proteins	33 ± 0.8	–	–

Table 7.2 Relative content of cytochromes (along the Soret band) and nucleic acids (according to the fluorescence of ethidium bromide) in fractions of bovine heart mitochondria

Fraction	D_{410}/C (per 1 mg)	F/C (%)
Mitochondria	0.29	92
PMC 0–0.45 μm	0.30	100
PMC 0–0.2 μm	0.30	96

Note: D_{410} is the optical density of the fraction (per 1 cm) in the Soret band, F is the fluorescence intensity of ethidium bromide, C is the relative concentration of the protein (optical density at 280 nm)

The concentration of ballast proteins in supernatants in all three fractions from bovine heart was almost the same. The oxygen-dependent oxidation of NADH by the respiratory chain was practically not detected in the supernatant, and only insignificant NADH: ferricyanide reductase activity (less than 0.09 μM NADH/min/mg protein) was present. Using the protein concentration in the fractions and in the supernatant obtained by centrifugation (taking into account the ballast protein), the ratio of particles of different sizes was calculated (Table 7.1).

7.1.2 *Fluorescent Properties*

The fluorescence properties of MCs and PMCs from rat liver were compared (Table 7.3). For this purpose, fluorescence of tryptophan residues of proteins was measured in the range of 300–400 nm, NADH (NADPH) at 400–500 nm and flavins at 500–580 nm.

Tryptophan fluorescence intensity was maximal in the total suspension of MCs. In the PMC fraction <0.22 μm, it was 85%, and in the PMC fraction <0.45 μm, it was 91%. The half-width of the spectrum varied slightly. These data indicate that fractions do not differ greatly in protein composition.

Table 7.3 Fluorescent properties of mitochondria and proto-mitochondria from rat liver

Fluorophore	Fraction	I_{max} (%)	λ_{max} (nm)	$\lambda/2$ (nm)
Tryptophan	MC	100	324 ± 2	60 ± 2
	PMC 0–0.22	85	324	59
	PMC 0–0.45	91	324	55
NADH	MC	98	430	90 ± 5
	PMC 0–0.22	100	430	90
	PMC 0–0.45	98	430	90
Flavins	MC	95	520	96
	PMC 0–0.22	100	520	96
	PMC 0–0.45	89	520	96

Note: I_{max} is the fluorescence intensity, λ_{max} is the maximum of the spectrum, $\lambda/2$ is the half-width of the spectrum

The fluorescence intensity of endogenous NADH (NADPH) in the MC and PMC fractions was almost the same. The half-width of the spectrum coincided in all cases. Since the main contribution to the fluorescence of endogenous NADH (NADPH) is made by molecules that are bound in the active centers of dehydrogenases [12], it can be concluded that PMCs contain approximately the same dehydrogenases as MCs.

The fluorescence intensity of endogenous flavins in the PMC fractions did not differ much from MCs. The half-width of the flavin spectrum in all samples was 96 nm.

The totality of the data obtained indicates the integrity of the protein structure of PMCs and the preservation of NADH dehydrogenase and flavin-containing proteins after these organelles pass through filters.

7.1.3 *Oxidation of NADH*

The NADH dehydrogenase complex contains several iron-sulfur clusters, flavin, as well as the associated NADH and ubiquinone [13–15]. This allows one to expect with high probability that PMCs may differ in the properties of this complex from that in MCs.

Under isotonic conditions, the intact inner membrane is impermeable for NADH [16]. However, if the MC fraction is placed in a hypotonic medium, rapid oxidation of NADH is observed. The maximum oxidation rate of NADH in MCs from bovine heart in 20 mM Tris buffer (pH 8.0) at 20 °C was ~0.17 μmol of NADH/min/mg protein. A similar picture occurred in PMC fractions (Fig. 7.2). In all fractions of PMCs and MCs from bovine heart, NADH oxidation was suppressed by respiratory inhibitors such as rotenone, antimycin, and cyanide or azide. The sensitivity of NADH oxidation to these inhibitors indicates the presence of respiratory chain in all fractions.

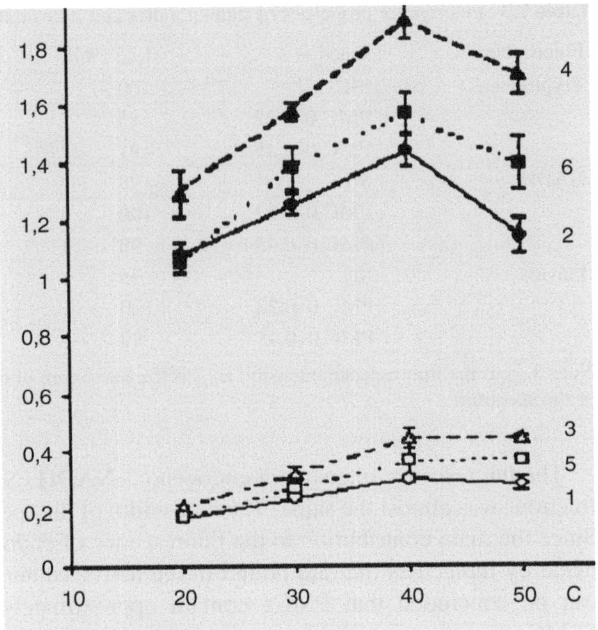

Fig. 7.2 The rate of NADH oxidation in respiratory chain (light symbols) and NADH: ferricyanide reductase reaction (dark symbols) depending on temperature in fractions of MCs (1,2), PMCs <0.45 (3,4) and PMCs <0.2 (5,6) from bovine heart

With the addition of ferricyanide, the NADH oxidation in all fractions increases sharply (Fig. 7.2), which corresponds to the known results obtained with MCs and purified NADH dehydrogenase [3, 13–15]. A quick reaction with ferricyanide is a shunt bypassing flavin, ubiquinone [9, 13] and, possibly, iron-sulfur clusters (the reaction is not inhibited by bathophenanthroline). The initial rapid reaction rate was 1 μmol/min/mg protein. Respiratory chain inhibitors did not affect this reaction in all fractions, as expected [13].

PMC fractions significantly differed in the speed and temperature dependence of NADH: ferricyanide reductase and NADH oxidase activities (Fig. 7.2). The dependence of activity on temperature for the smallest particles is sharper. This means that there are differences in the composition of the PMC and MC membranes. Small particles are more active.

MCs and PMCs isolated from rat liver maintained good integrity. Isolation of these organelles from bovine heart is associated with mechanical stresses on the tissue. It is very likely that fractions of small particles obtained from the heart consist not only of PMCs, but also of MC fragments. On the other hand, the difference in the temperature dependences of the fractions (Fig. 7.3) is noteworthy, which indicates qualitative differences in the respiratory chain of particles of different sizes. In addition, the change in the rate of both reactions at 50 °C compared with the speed at 40 °C is more pronounced in large particles. For example, when the temperature rises to 50 °C, the oxidation rate of NADH in the respiratory chain of the MC fraction decreases, does not change in the 0.45 micron fraction, and slightly increases in the 0.2 micron fraction.

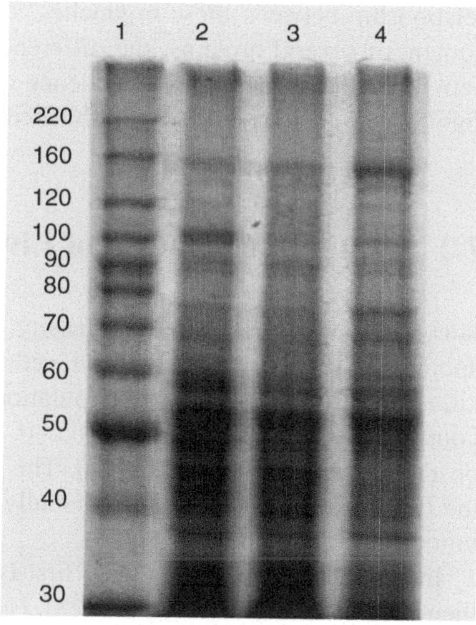

Fig. 7.3 Protein profile of mitochondrial fractions in SDS-PAAG. 1 – molecular weight markers. 2 – total suspension of MCs. 3 – PMCs of size 0–0.45 microns. 4 – PMCs 0–0.22 microns. Cardiac MCs and PMCs were treated with SDS and applied to the gel at a concentration of 120 μg/ml

7.1.4 Effect of Sucrose on NADH: Ferricyanide Reaction

The NADH oxidation rate of in the respiratory chain depends on the osmoticity of the medium [13, 16]. The speed of the shunting NADH: ferricyanide reductase reaction in a hypotonic medium (without sucrose) increases. This can be explained by membrane damage and partial fragmentation of the NADH oxidase complex [13]. Small PMCs in terms of NADH: ferricyanide reductase reaction rate is similar to MCs, but the dependence on sucrose concentration is much stronger. This indicates the similarity, but not the identity of the properties of the membranes with respect to the tonicity of the medium.

7.1.5 Protein Profile

The protein profile of the PMC and MC fractions from bovine heart (electrophoresis under standard conditions in the presence of SDS and a reducing agent; Fig. 7.3) differed in the intensity of several bands. In particular, in the MC fraction, the content of proteins with a molecular weight of 103 and 58 kDa is noticeably higher than in PMC fractions, and the protein content of 155 kDa is increased in the lightest PMC fraction.

Conclusion The data obtained indicate a qualitative similarity between PMCs and MCs (similar protein composition, similar microenvironment of tryptophan residues of proteins, close contents of NADH, flavins and cytochromes, the presence of succinate and NADH dehydrogenase activities, etc.), which indicates a close relationship between these organelles. At the same time, some characteristics (the content of several proteins, the difference in succinate and NADH dehydrogenase activities, the temperature dependence of the NADH oxidase reaction, etc.) differ significantly. The results were published in [18].

7.2 Proto-mitochondria of Liver Cells

Intracellular MCs differ in size, density, participation in cell metabolism, aging and apoptosis [19–23]. In specialized cells of animal organs (liver, muscles, heart, kidneys, etc.) there are three populations of these organelles (at the same time): young PMCs with a diameter of 0.2–0.45 μm, mature MCs – about 1 μm, and old postMCs – about 1–2 μm [24, 25]. The ratio of these three populations depends on the type of cells, age, etc. [24]. Usually, PMCs in cells make up about 30% of all mitochondrial organelles [24, 25].

In our laboratory, PMCs were first isolated from mammalian cells and some of their properties were studied [25–28]. Using electron microscopy, it was determined that PMC 0.45 microns in size has both outer and inner membranes. High activities of NADH: ferricyanide reductase, NADH: tetrazolium reductase, succinate: ferricyanide reductase and succinate: tetrazolium reductase in PMCs were detected. Moreover, the number of cytochromes in PMCs is lower than in MCs. A group of authors [29] studied the properties of plant PMCs in cell seedlings.

It was interesting to carry out the study of biochemical properties of PMCs in comparison with mature MCs. The total MC fraction from rat liver was isolated according to the standard method with modifications [25–27], which made it possible to obtain a wide range of heavy and light organelles. The liver was placed in 20 ml of medium containing 0.25 M sucrose, 0.5 mM EGTA and 10 mM HEPES (pH 7.4) at 4 °C, after which it was pressed through a press and then homogenized. The homogenate was centrifuged for 10 min at 1000 g. The supernatant was centrifuged for 20 min at 10,000 g, as a result of which the precipitate contained not only MCs, but many lighter PMCs. The precipitate was suspended in 10 ml of isolation medium, aliquoted and used for experiments on the same day, and also frozen at −18 °C for subsequent use. Protein concentration was determined according to [30].

The MC suspension (at a protein concentration of about 1 mg/ml) was passed through Whatman millipore filters on a nitrocellulose base with a pore size of 0.22 and 0.45 microns or through microporous capron membrane filters of NPP Technofilter with pore sizes of 0.15, 0.25, 0.45, 0.8 and 2.0 microns. Calibrated PMCs were isolated by filtration with a diameter of (a) not more than 0.15 μm, (b) not more than 0.25 μm, and (c) not more than 0.45 μm were used for comparison with MCs of sizes over 0.8 μm, as well as with the total MC fraction.

The respiratory activity of PMCs and MCs was determined by the rate of oxygen consumption using the Clarke electrode. The respiratory control was calculated as the ratio of the rate on succinate to the rate after adding 200 μM ADP. The incubation medium contained 5 mM succinate, 2 mM potassium phosphate and 150 mM sucrose, pH = 7.5.

50 μl of the initial suspension of PMCs or MCs were mixture with 100 μl of incubation buffer (pH 7.4) and 50 μl of a solution of 0.1 M NaCl. The suspension was gently shaken on a vortex for 1 min to disintegrate the aggregated MCs, and then it passed through a millipore filter with pore sizes of 0.15, 0.25, 0.45, 0.8, or 2.0 μm. The size of the particles was determined by phase contrast and fluorescence microscopy.

Acridine orange fluorescent dye was used for staining, followed by observation with a FluoView FV10i-w confocal microscope. 15 μl of dye was added to 200 μl of the resulting suspension (initial concentration 0.1 mg/ml). Organelles were stained for 10–20 min. Samples were centrifuged for 3 min at 7000 rpm and then washed from excess dye. 100 μl of buffer was added to the precipitate, passed through a millipore filter and observed in a fluorescence mode on a confocal microscope.

Fluorescence spectra of flavins, tryptophan residues of proteins and acridine orange were recorded on a Perkin-Elmer MF44B.

The amount of DNA was determined in three ways: (a) by optical density at 250 nm on a spectrophotometer, (b) by the fluorescence intensity of acridine orange (excitation 450 nm, radiation 600 nm), (c) electrophoretically in a 1% agarose gel. In the last case, DNA from PMCs and MCs was isolated accordingly to [31].

The protein composition of PMCs and MCs was compared by denaturing gel electrophoresis in 10% SDS. To prepare the samples, the total suspension of MCs was passed through millipore filters, after which these filters were thoroughly washed with 0.6 ml of working buffer containing 2% SDS. Samples obtained by washing off the filters were then analyzed using the Laemmli method [32, 33]. 25 μl of a PMCs or MCs (100 μg in the sample) were applied to each lane.

The protein and enzymatic composition of PMCs (the size up to 0.45 μm) and MCs (0.8–2 μm) were also determined by gel filtration on a Sephadex G-50, G-75 or G-100 column or Toyopearl hw 55. 10 mM Tris-HCl was used as buffer (pH 7.5) with 50 mM NaCl and 0.1–3% SDS or dodecyl-β-D-maltoside, or 0.1% α-octyl-β-D-glucopyranoside.

The enzymatic oxidation reaction of succinate or NADH in PMCs and MCs was detected at 540 nm by the optical density of formazan formed from pNTV, according to [24]. 25 μl of 10 mM pNTV and 10 μl of 10 mM NADH or 50 μl of 10 mM succinate were added to 500 μl of PMCs, MCs or an enzyme solution in a working buffer (containing 0.1% detergent). Both the PMC and MC activities were recorded at points for 20 min at 30 °C, and for PMCs and MCs, destroyed with a detergent, 16 h at 20 °C.

The total suspension was passed through filters with a pore size of 0.8; 0.45 and 0.22 microns to obtain fractions of MCs and PMCs. Phase contrast microscopy showed that the initial preparation contained 65% organelles with a diameter of 0.45–2 microns and 35% with a diameter of less than 0.45 microns (Fig. 7.4).

To confirm the mitochondrial nature of the resulting organelles, the preparations were treated with a 0.1% solution of orange acridine, which is able to stain DNA. In principle, this dye can stain not only DNA, but it is in DNA that it fluoresces most

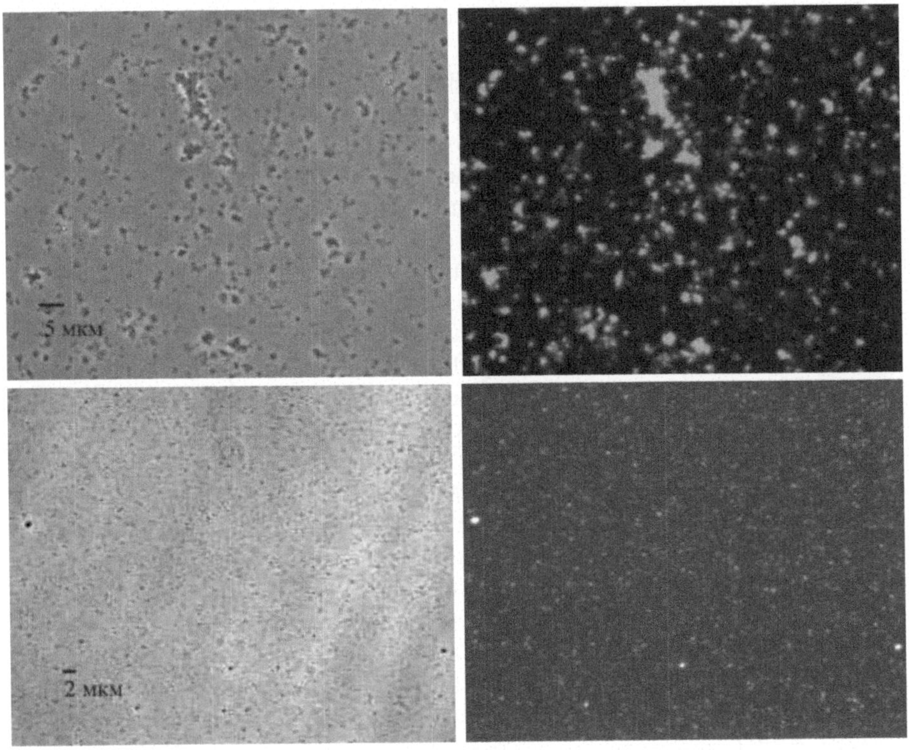

Fig. 7.4 Mitochondria (above) and proto-mitochondria <0.45 microns (below) obtained from rat liver. Confocal microscopy with FluoView FV10i-w microscope (Olympus) in phase contrast mode (left) and green fluorescence when stained with acridine orange (right). Excitation laser was 473 nm

strongly if there is a lot of DNA. When staining of MCs, noticeable dye fluorescence was observed. PMCs were stained much less intensely (Fig. 7.4). This is consistent with the data [25, 28] that the number of DNA copies in each PMC is less than in each MC. The signal from acridine-orange from PMCs <0.22 micron was weaker than from PMCs <0.45.

Another proof of the mitochondrial nature of organelles was the determination of the presence of DNA in them by electrophoresis. DNA isolated from PMCs and MCs was located in the gel at the same level (Fig. 7.5), higher than low molecular weight oligonucleotide standards and corresponds to a mass of 16 kbs. Moreover, the number of copies of DNA molecules in PMC is much smaller than in MC.

Table 7.4 shows spectrophotometric data to evaluate the protein/DNA ratio in various fractions of PMCs. These data confirm previous observations [25, 27, 28] that the amount of DNA in PMC when it matures to MC increases.

To make a more complete comparison of the protein composition, the steps of precipitation of proteins by trichloracetic acid and washing by acetone before gel electrophoresis were excluded, since both the hydrophobic and hydrophilic proteins are part of MCs and PMCs (although it somewhat worsened the quality of

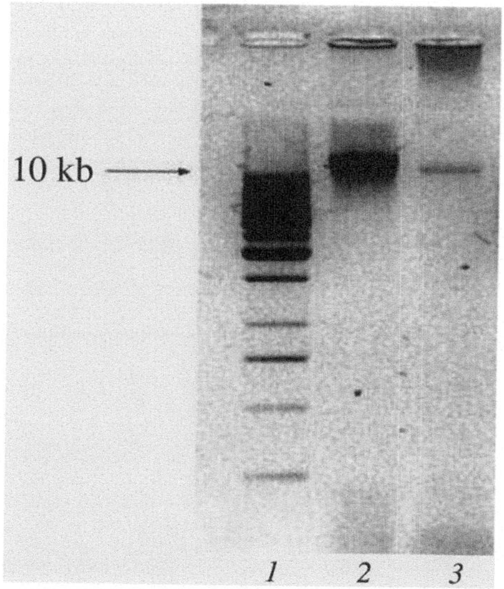

Fig. 7.5 Gel electrophoresis of DNA from MCs (particle diameter over 0.8 microns, lane 2) and PMCs (diameter 0.25–0.45 microns, lane 3). Markers – lane 1

Table 7.4 Spectrophotometric ratio of protein/DNA in mitochondrial fractions from rat liver

Fraction	D at 286 nm (proteins)	D at 250 nm (DNA)	D286/D250
MC	0.55	0.75	0.74
MC < 0.8 micron	0.63	0.88	0.71
PMC < 0.45 micron	0.61	0.85	0.71
PMC < 0.15 micron	0.56	0.76	0.74

Note: It was measured after organelle destruction in 2% SDS. Protein concentrations did not equalize

electrophoresis). The complete list of MitoCarta contains about 1100 gene loci encoding MC proteins [24]. One of the properties of the MC proteome is a large scatter in the ratios of its components [35]. The discrepancy in the amounts of different MC proteome proteins in different objects and in different conditions is multiple [36]. Apparently, this is connected not only with the fact that 15–25% of MC proteins are encoded by their own DNA and 75–85% by nuclear DNA, but also with the strong variability of the MC genome [34].

Nevertheless, the protein composition of PMCs (particles with diameter from 0.22 to 0.45 μm) and MCs (particles over 0.8 μm) from rat liver is almost the same (Fig. 7.6). In both cases, two major bands are visible at the standard level of 55–60 kDa, and a band of similar intensity at the level of 30 kDa. Some quantitative differences exist for proteins with a weight of 45 and 50 kDa, the intensity of the bands of which in MCs is twice as high as in PMCs. The obtained data on rat liver PMCs are somewhat different from those that were identified for PMCs from bovine heart [25].

A comparative analysis of the protein composition of PMCs and MCs was also carried out using gel filtration. For both PMCs (diameter from 0.22 to 0.45 μm) and

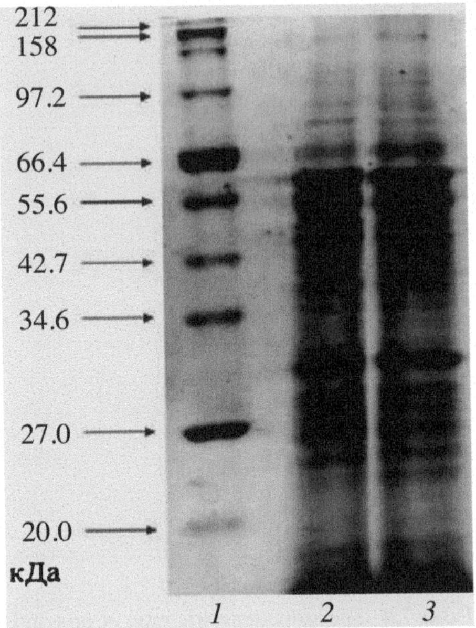

Fig. 7.6 Electrophoresis of proteins of MCs and PMCs in a 10% polyacrylamide gel in the presence of 0.1% SDS. 1 – standards, 2 – MCs > 0.8 microns, 3 – PMCs <0.45 microns

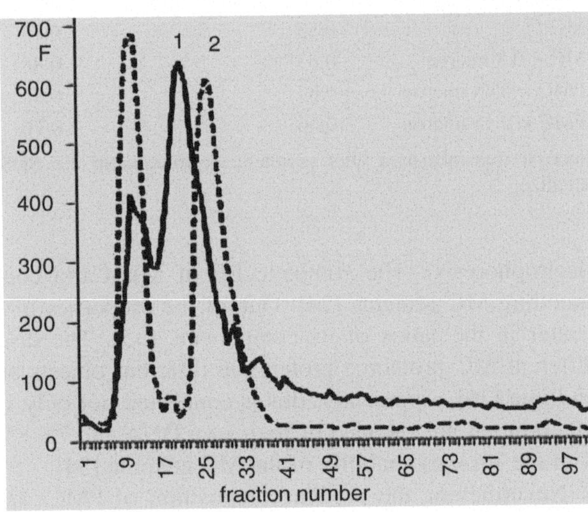

Fig. 7.7a Gel filtration of mitochondrial proteins (0.8–2 microns) on a Sephadex G-75 column. Fluorescence of tryptophans (1), fluorescence of flavins (2)

MCs (diameter over 0.8 μm), two high molecular weight protein peaks were found, each of which has NADH dehydrogenase and succinate-dehydrogenase activities, as well as flavin fluorescence is detected.

Figure 7.7a and 7.7b shows a typical result of gel filtration analysis of the MC and PMC proteins on a Sephadex G-75 column. The amount of flavoproteins in PMCs, judging by gel filtration, is noticeably lower than in MCs. Interestingly, the profile of PMCs is poorer, it differs markedly from MCs.

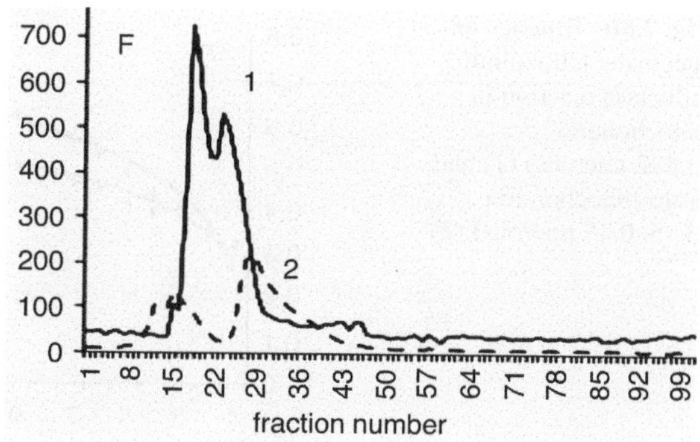

Fig. 7.7b Gel filtration of proto-mitochondrial proteins (0.15–0.45 microns) on a Sephadex G-75 column. Fluorescence of tryptophans (1), fluorescence of flavins (2)

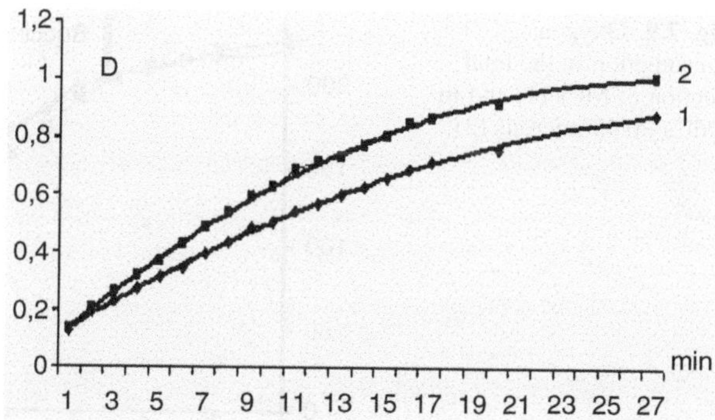

Fig. 7.8a Kinetics of NADH: tetrazolium reductase reaction in mitochondria 0.8–2 microns (1) and proto-mitochondria 0.15–0.45 microns (2)

PMCs possess a high NADH dehydrogenase activity, and NADH: ferricyanide reductase activity of small (0.1 and 0.2 μm) PMCs is even higher than larger ones [25, 28]. PMCs have high NADH: tetrazolium – reductase activity. In Fig. 7.8a shows the kinetics of the reaction when using pNTV. The speed in the case of PMC is 25% higher than MC. Therefore, in PMCs, there are a certain number of NADH dehydrogenase subunits, which are sufficient for a shunt reaction with tetrazolium. Judging by the rates, the availability of active centers in PMC for NADH and tetrazolium is higher than in MC.

Succinate: tetrazolium reductase activity of PMCs was, conversely, 30% lower than MCs (Fig. 7.8b). Although there are some succinate dehydrogenase subunits in PMCs that are sufficient for a shunt reaction with tetrazolium, the number of enzyme molecules and/or the availability of the active site for succinate and tetrazolium are lower than in MCs.

One of the most important functional characteristics of MCs is respiration – oxygen consumption. The specific rates of succinate oxidation in the respiratory chain of PMCs and MCs, as well as the respiratory control value upon addition of ADP, were determined polarographically. As can be seen from Fig. 7.9, PMCs <0.45 microns respires on succinate similar to MCs (also – Table 7.5).

Very low respiratory control in small PMCs <0.15 micron means that either the respiratory chain is disconnected from ATP synthetase, or that ATP synthetase itself

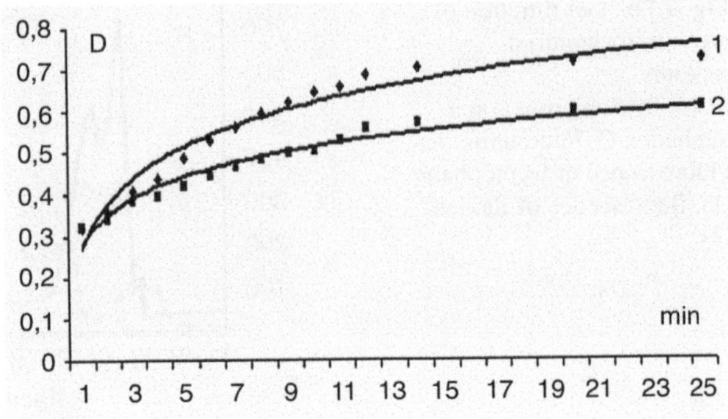

Fig. 7.8b Kinetics of succinate: tetrazolium reductase reaction in mitochondria (0.8–2 microns) (1) and proto-mitochondria (0.15–0.45 microns) (2)

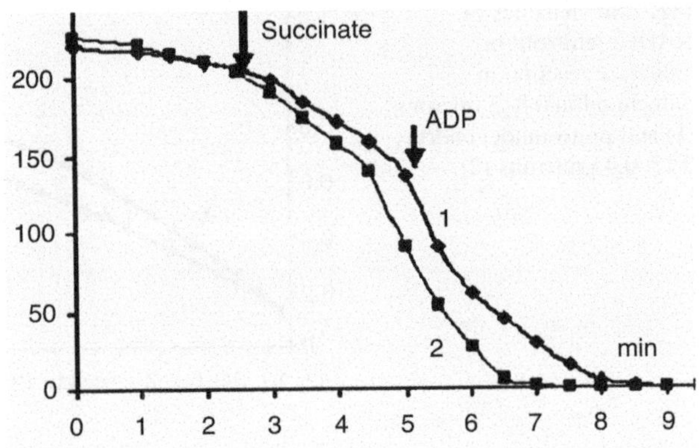

Fig. 7.9 Oxygen consumption in the total fraction of MCs (1) and in PMCs <0.45 microns (2)

Table 7.5 Respiratory activity in mitochondrial fractions from rat liver

Fraction	Succinate-dependent respiration rate (μM O_2/min/mg)	Respiratory control	NADH-dependent respiration rate (μM O_2/min/mg)
MC	0.8	1.9	0.3
PMC < 0.15	1.6	1	0.1
PMC < 0.45	1.2	1.5	0.2
PMC < 0.45 with 10 mM malonate	0.1	–	–
PMC < 0.45 with 5 μM antimycin	0	–	–
PMC < 0.45 with 5 μM rotenone	–	–	0

Note: the concentration in all samples was ~1 mg/ml

is absent. The PMC respiration on succinate was competitively inhibited by malonate, as well as respiratory chain inhibitors – antimycin and azide. The result of the high succinate oxidase activity of PMCs is not completely consistent with the data of Lanny et al. [37] that smaller MCs have a low respiration rate, but are

Table 7.6 The effect of para-nitrotetrazolium violet on the respiration of frozen-thawed MCs and PMCs

Organelles	Respiratory rate (μM O_2/min) at 10 mM succinate		Respiratory rate (μM O_2/min) at 0.2 mM NADH	
		+ 0.5 mM pNTV		+ 0.5 mM pNTV
Mitochondria	23	23	28	4
Proto-mitochondria	15	13	8	2

consistent in that they have a lower respiratory control. It should be noted that it was suggested in [37] that small MCs are young. A. Belyakovich [24] proved it much earlier.

In PMCs, respiration is observed due to the oxidation of NADH, which is competitively suppressed by an excess of NAD. Moreover, the use of NAD**P**H instead of NADH did not lead to respiration. Respiration during oxidation of NADH is blocked by rotenone (Table 7.5).

The rate of respiration during oxidation of added substrates (10 mM succinate or 0.2 mM NADH) in the case of MCs damaged by freezing was higher than PMCs.

Interestingly, pNTV suppressed the PMC respiration during NADH oxidation to much greater extent than MCs damaged by freezing (Table 7.6).

The inhibitory effect of pNTV on respiration is apparently due to that a part of electrons goes to this acceptor, instead of going through respiratory chain to oxygen. The respiration of thawed MCs on succinate was not suppressed at all by tetrazolium. In the case of PMCs, it decreased slightly. Possibly, the greater effect of pNTV on the PMC respiration than MCs is associated with a higher availability of active centers.

So, PMCs have an active respiratory chain. Based on the fact that the smallest PMCs contain few flavins and cytochromes, we can assume that their chain is simpler: electrons from NADH dehydrogenase can go either (a) directly to cytochrome oxidase, or (b) directly to oxygen.

Based on all the above facts, confirming the data [24–29] about the existence of germinal and young PMCs in specialized mammalian cells and the fundamental commonality of their basic properties with the properties of mature MCs, it can be suggested that germinal PMCs can occur in cells due to DNA 16 kbs entering the cytoplasm during aging and degradation of MCs (Fig. 7.10). In the process of the cell cycle, the PMCs can develop into adult organelles – MCs. After their aging, the formation of large old post-MCs [24]) take place. The subsequent lysis of these organelles provide by proteases and lysosomes. It leads to the release of DNA into the cytoplasm. Aging of MCs is accompanied by the accumulation of lipofuscin and a decrease in the amount of phospholipids; phospholipase A2 plays an important role in this [38].

Conclusion A number of properties of proto-mitochondria, small young mitochondrial organelles of animal cells, were studied. PMCs of 0.45 microns in diameter have a complete respiratory chain. No qualitative differences of major proteins were

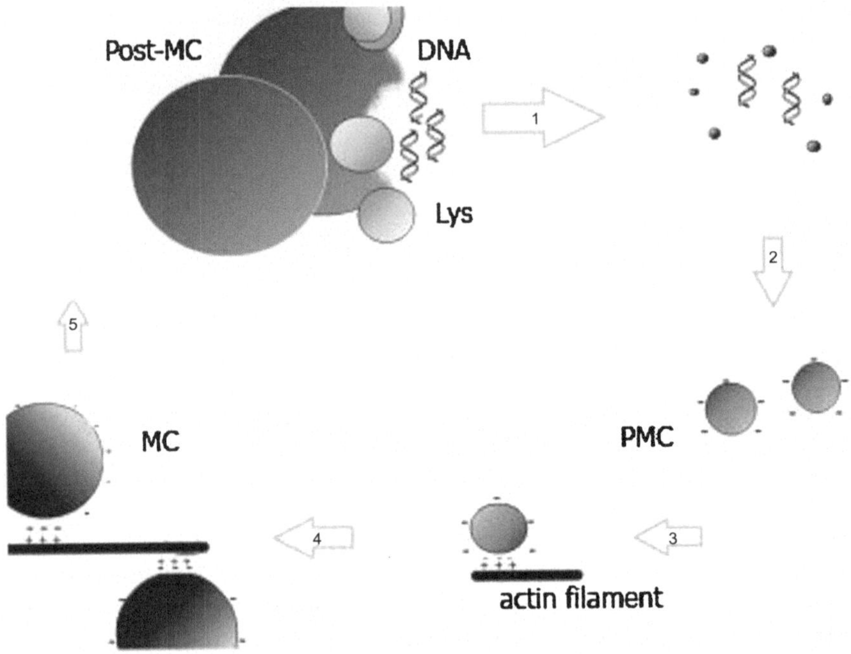

Fig. 7.10 Mitochondrial cycle in specialized animal cell. *1* – destruction of old post-mitochondria by proteases and lysosomes lead to release of DNA and proteins into the cell cytoplasm, *2* – the formation and growth of proto-mitochondria, *3* – the interaction of proto-mitochondria with actin filaments, *4* – maturation of proto-mitochondria to mitochondria, *5* – aging of mitochondria and the formation of post-mitochondria

found in PMCs with a diameter of 0.25–0.45 μm and MCs, but some quantitative differences were revealed. The indicated differences, apparently, reflect the process of intracellular maturation of PMCs to MCs. The results were published in [39].

7.3 Germinal Proto-mitochondria

One of the main functions of mature MCs is the generation of a transmembrane potential, accompanied by the synthesis of ATP [40]. MCs of different cells differ in size, density, participation in metabolism, aging and apoptosis [41–44]. In the cells of organs (liver, muscles, heart, kidneys, etc.), MCs practically do not divide, but appear presumably in the form of germinal PMCs with a diameter of less than 0.2 microns, which gradually grow to mature MCs with a diameter of about 1 micron, eventually degrading to old post-MCs with a diameter of more than 1 micron [45, 46]. The ratio of populations of PMCs, MCs, and post-MCs in each cell varies greatly [45]. PMCs <0.45 microns in specialized cells make up about 30% of all mitochondrial organelles [45, 46].

PMCs were first isolated in our laboratory from cells of animal organs (heart and liver) and some of their properties were studied [46–52]. On the example of hepatic PMCs with a diameter of 0.45 microns, it was shown that these spherical particles are not scraps of MCs. It is important that they have respiration and some phosphorylating activity. PMCs with a diameter of 0.25–0.45 microns have a complete respiratory chain, as evidenced by the consumption of oxygen on cytochrome oxidase during the oxidation of succinate and NADH. In addition, PMCs have high succinate- and NADH-tetrazolium reductase activities. Horizontal electrophoresis showed that PMCs 0.25–0.45 microns contain 16 kbs monomeric DNA (in the MC, in addition to the 16 kbs, there is a minor satellite DNA). Gel filtration and vertical electrophoresis showed the coincidence of most protein bands of PMCs 0.25–0.45 and MCs, but with some quantitative differences. The spectrum of protein tryptophan fluorescence of PMCs 0.25–0.45 almost does not differ from MCs, but the tyrosine component is slightly increased.

It was interesting to study a number of properties of the smallest (germinal) PMCs with a diameter of less than 0.2 microns. This is necessary to understand the mechanism of their occurrence and maturation to full-fledged MCs.

A total MC fraction from the liver of Wistar rats was isolated according to the usual method with modifications [46–48], which made it possible to obtain a wide range of large heavy and small light organelles. The work was carried out at 4 °C. All solutions were cooled on ice. The liver of a 4-month-old rat was placed in 60 ml isolation medium containing 250 mM sucrose, 1 mM EDTA and 10 mM HEPES (pH 7.5), and then pressed through a press and homogenized. The homogenate was centrifuged for 10 min at 1000 rpm. The supernatant was centrifuged for 15 min at 3300 rpm to obtain the sediment of heavy MCs, which was diluted with isolation medium and packaged in Eppendorf tubes. The supernatant was centrifuged for 20 min at 5600 rpm. The precipitate (light fraction) was suspended in 6 ml of the same isolation medium and immediately passed through Whatman nitrocellulose millipore membranes with 0.2 μm pores or 0.22 μm millipore PVDF filters. The obtained PMC filtrate was aliquoted and used for experiments. Before spectrophotometric and fluorescence experiments, 0.5% SDS was added to dissolve the particles and minimize light scattering.

For comparison, the heavy fraction of MCs or calibrated MCs was taken. The calibrated MCs were obtained by filtering a diluted suspension of heavy MC fraction through 1-μm millipore filters. In this case, MCs were collected from these filters by soaking in 0.5% SDS.

The destruction of PMCs and MCs by SDS eliminates light scattering, i.e. it made the turbid suspension to completely transparent solution. The residual small contribution of light scattering in the region of the bands in the UV and visible region was removed by approximation from 700 nm, where there is no light absorption, but only light scattering.

Absorption spectra in the UV region were recorded on Carry Eclipse or M-40 in a special compartment. Fluorescence spectra were recorded in 1-cm quartz cuvettes on MF44B Perkin-Elmer and Carry Eclipse using transparent detergent solutions obtained from PMCs and MCs by SDS (in the absence of potassium ions to minimize detergent aggregation). The total protein content of PMCs and MCs in spectroscopic experiments was 0.35 mg/ml. Protein concentration was determined by the UV method [53].

Optical microscopy of PMCs and MCs was performed on LUMAM I-2 and Axio Imager.Z1 microscopes in a transmitted, phase-contrast, and fluorescence mode. For staining DNA in PMCs and MCs, a Hoechst 33342 fluorescent dye was used.

IR absorption spectra of the PMC and MC films were recorded on FT-801 Fourier transform infrared spectrometer. In this study, MCs and PMCs were isolated from the liver in the presence of 300 mM sodium chloride without sucrose, without EDTA, and without HEPES absorbing in the IR region. Films were prepared by drying of suspensions on calcium fluoride plates, transparent in IR. By the way, the MC and PMC suspensions respired well on succinate.

The protein composition of PMCs and MCs was compared by denaturing gel electrophoresis in 10% SDS. 25 μl of PMCs or MCs (100 μg of protein in a sample) was applied to the lane.

The respiratory activity of PMCs and MCs was determined by the rate of oxygen consumption using the Clarke electrode. The incubation medium contained 5 mM succinate, 2 mM potassium phosphate and 150 mM sucrose, (pH 7.5).

PMCs <0.2 μm are not visible under optical microscopy either in transmitted light, or in the phase-contrast mode, or in the fluorescence mode (not shown in a figure). This is due to the fact that (a) the particle size is significantly smaller than the light wavelength, (b) the refractive index of the particles is not high, (c) optical density of each particle is negligible. However, when stained with Hoechst (a specific dye for DNA), they become visible in phase-contrast and, especially, fluorescent modes (Fig. 7.11).

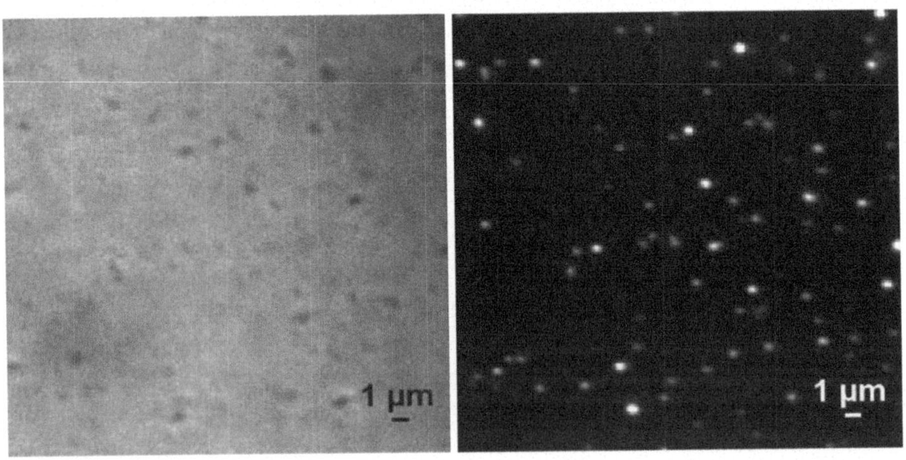

Fig. 7.11 Optical microscopy of PMCs from rat liver, obtained by filtration of the light fraction of MCs through a millipore filter with a pore diameter of 0.2 microns, with Hoechst staining in the "bright field, phase contrast" mode (left) and "fluorescence" (right; excitation – 350 nm; interference filter). Axio Imager.Z1 microscope, EC Plan-Neofluar 100x/1.30 Oil Iris M27 lens

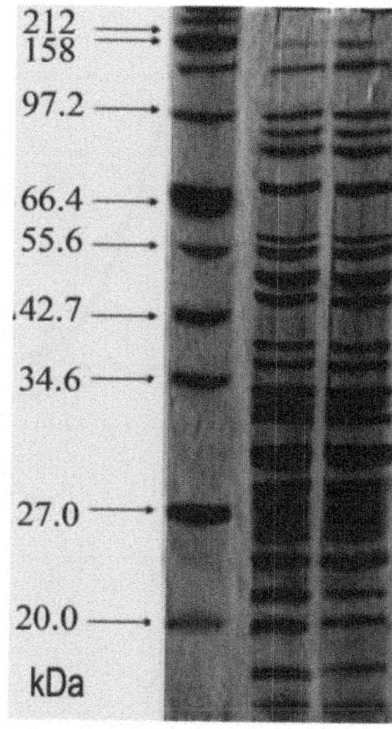

Fig. 7.12 Protein electrophoresis of PMCs <0.22 microns (on the right) and MCs (in the center) in 10% polyacrylamide gel in the presence of 0.1% SDS in comparison with the marker standard (on the left)

In fluorescence mode, the picture is much sharper than in phase contrast. This is due to that a shorter-wavelength illumination of 350 nm is used in the fluorescence mode. The diameter of some particles in Fig. 7.11 seems to be larger than 0.2 microns, due to that they were out of focus, and also due to diffraction and dispersion. Since the Hoechst dye specifically stains DNA, it means that PMCs <0.2 μm contain the DNA molecules. This result is consistent with the data of electrophoresis [50] on the presence of DNA in the PMC 0.25–0.45 microns.

A comparative analysis of the total protein composition of PMCs and MCs using gel electrophoresis in 10% PAGE in the presence of 0.1% SDS did not reveal any significant differences between the major proteins of germinal PMCs and matured MCs (Fig. 7.12), which proves the complete relationship of these organelles. There is a quantitative difference only for low molecular weight proteins <22 kDa, the content of which in PMCs is smaller than in MCs. From electrophoresis it also follows that the fraction of germinal PMCs, if it contains any protein impurities, they do not noticeably differ from impurities in the MC fraction.

For functional proteins, contained in minor quantities, the differences between germinal PMCs and mature MCs are quite significant. Judging by the characteristic absorption spectrum in the visible region, PMCs <0.2 microns have fewer cytochromes than MCs (weakened Soret band; the long-wavelength part of the band is shown at 480 nm in Fig. 7.13), and they do not show cytochrome oxidase (at 650 nm).

Thus, PMCs <0.2 microns do not have formed respiratory centers. That is why, unlike PMCs with a diameter of 0.25–0.45 μm, which have respiration and even respiratory control [46, 50], PMCs <0.2 μm either hardly respire or uptake oxygen at NADH oxidase (it depends on many conditions).

Judging by the intensity of flavin fluorescence (maximum at 525 nm), the total content of flavoproteins in germinal PMCs, on average, is 35% (±8%) less than in

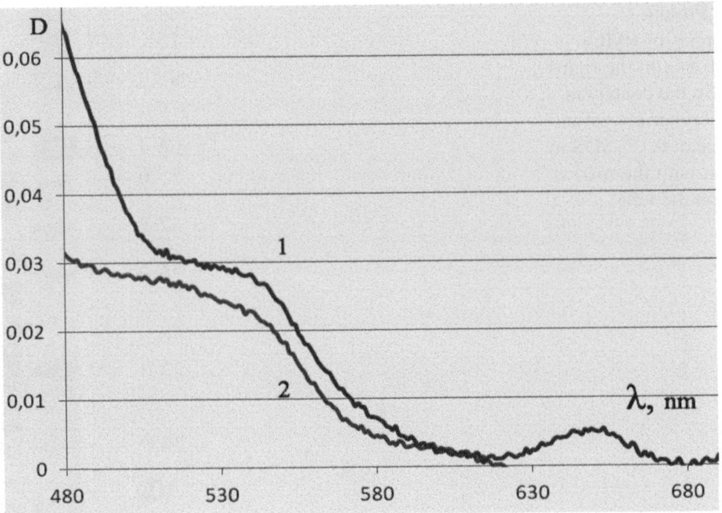

Fig. 7.13 Absorption spectra of cytochromes of MCs (1) and PMCs <0.2 microns (2) after the destruction of particles by detergent (0.5% SDS). Characteristic spectra are given, excluding light scattering

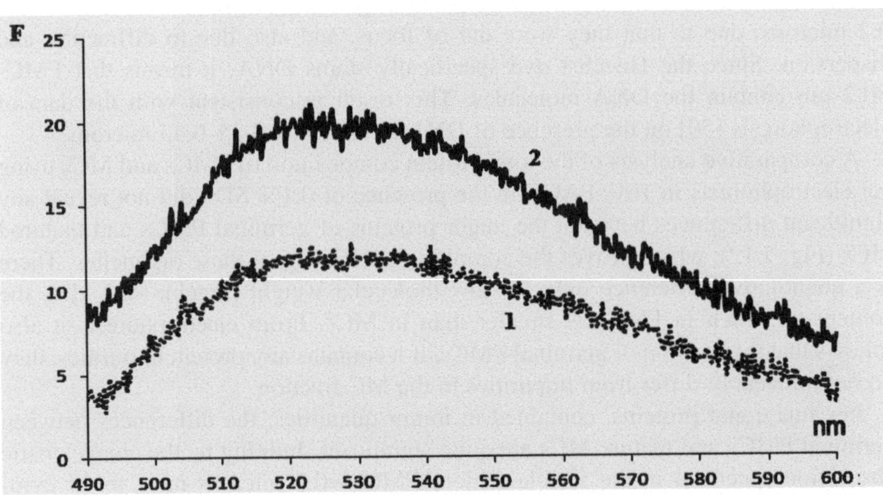

Fig. 7.14 Flavin fluorescence spectra (excitation – 450 nm) PMCs <0.2 microns (1) and MCs (2) after the destruction of suspension particles with detergent (0.5% SDS)

MCs (Fig. 7.14). This is the expected result, if we assume that during the maturation of PMCs, the number of flavoproteins should increase in order to ensure the functioning of the respiratory chain. In hepatic MCs there is a flavin-containing NADH dehydrogenase and several FAD-containing flavoproteins [54]. In germinal PMCs there are fewer of them than in MCs.

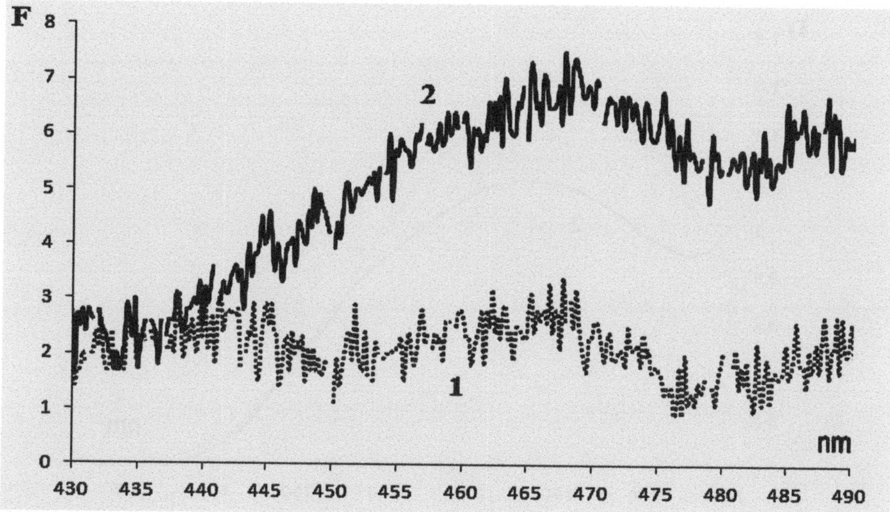

Fig. 7.15 Lipofuscin fluorescence spectra (excitation – 360 nm) of PMCs <0.2 μm (1) and MCs (2) after the destruction of suspension particles with detergent (0.5% SDS). The contribution from the flavin fluorescence is visible at 490 nm

What is especially important, the germinal PMCs practically do not contain the "aging pigment" – lipofuscin: the fluorescence spectrum does not have a characteristic "blue" band, while in the MCs it is quite noticeable (Fig. 7.15).

In various experiments on different adult rats, the intensity of the lipofuscin band in germinal PMCs was 3–5 times lower than in MCs. This fact means that PMCs <0.2 microns are very young organelles, in which, unlike MCs, Schiff bases due to lipid per-oxidation and protein-to-protein cross-linking have not yet formed. Interestingly, in the monograph [45], a strong lipofuscin fluorescence of post-MCs was described. Lipofuscin is formed in aging MCs without any participation of lysosomes [51], and the appearance of lipofuscin can be activated by moderate heating. It should be noted that the band at 465 nm in our conditions does not belong to endogenous NADH and NADPH, because MCs were initially incubated.

The UV absorption spectrum of germinal PMCs is more likely DNA than protein: the maximum is at 258 nm – in the shorter wavelength region than the MC spectrum, where the contribution of DNA to UV absorption is also significant (Fig. 7.16). A small tryptophan shoulder of proteins at 288 nm is clearly manifested in the spectrum. The difference spectrum of UV absorption of PMCs and MCs approximately corresponds to the DNA band. Germinal PMCs have less total protein than MCs and therefore have a higher spectral ratio of DNA/protein. The presence of a pronounced DNA absorption band in the PMC spectrum once again confirms that the germinal PMC fraction is not too contaminated with microsomes or other particles. Otherwise, the spectrum would not be as DNA, but protein.

When comparing the absorption spectra of PMCs <0.2 μm and MCs in the middle IR region (Fig. 7.17) with the IR spectra of marker proteins (BSA, etc.), with pure

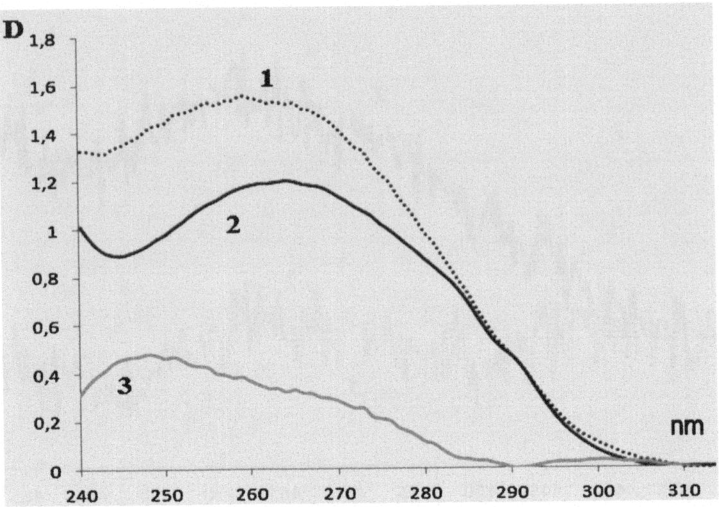

Fig. 7.16 UV absorption spectra of PMCs <0.2 microns (1) and MCs (2) after the destruction of suspension particles by detergent (0.5% SDS), as well as the difference spectrum (3)

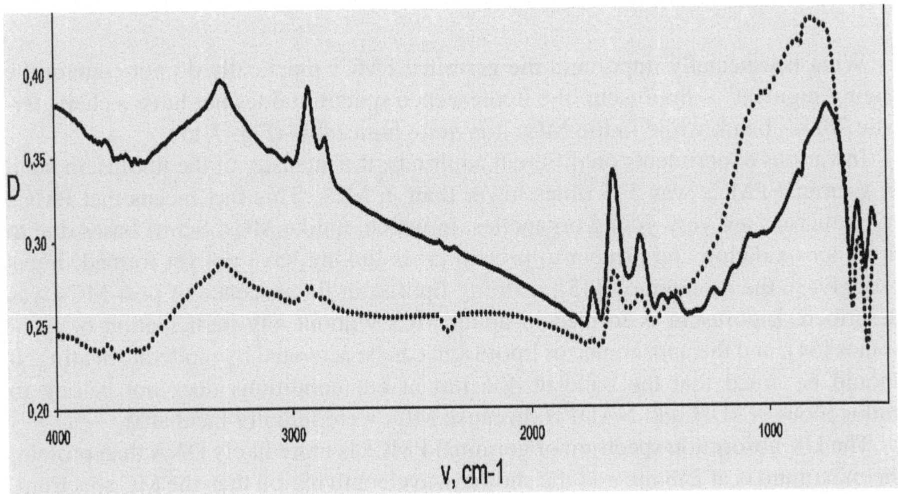

Fig. 7.17 IR absorption spectra of dried PMC films (···) and MC films (-)

DNA and with lipids (their spectra are not given here), it was found that the main contribution to IR absorption PMCs and MCs, as expected, give proteins. The set of intense bands at 3293, 2924, 1652, and 1542 cm^{-1} in MCs and PMCs belongs to them. A small band at 3786 cm^{-1} belongs to purines and pyrimidines of DNA [55]. Duplet 640–610 cm^{-1} belongs to DNA phosphates, because a similar doublet is present in the spectrum of sodium phosphate. A broad band with maxima in the region of 750–800 cm^{-1} belongs to lipids and other components.

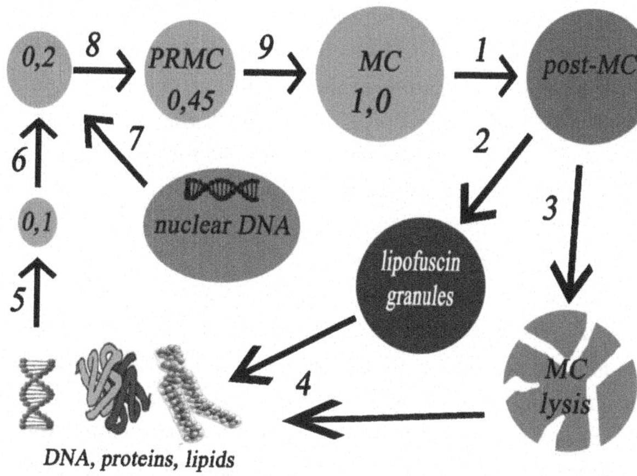

Fig. 7.18 Scheme of the MC life cycle in a specialized animal cell: 1 – aging of MC to post-MC, 2 – degradation of post-MC into lipofuscin granules or 3 – lysis of post-MC by proteases and lysosomes, 4 – DNA output (with some proteins) into cytoplasm, 5 – "nucleation" of the 0.1-micron PMC, 6 – synthesis of a number of membrane proteins, 7 – cooperation with nuclear DNA, synthesis of most proteins and membrane formation, 8 – growth of PMC, 9 – further growth and development of PMC to MC

The peak at 3293 cm^{-1} is due to stretching vibrations of the NH and OH groups, and the peak at 2924 cm^{-1} is due to stretching vibrations of the OH and CH groups [55]. The peak at 1652 cm^{-1} is due to stretching vibrations of CO groups, as well as deformation vibrations of OH groups of bound water remaining after drying of PMCs and MCs. The peak 1542 cm^{-1} is due to deformation vibrations of the NH groups.

The IR spectrum of germinal PMCs (with a diameter less than 0.2 μm) is very similar to the MC spectrum, which once again confirms the close relationship between PMCs and MCs. However, the intensities of the bands differ markedly. The position of the peaks does not completely coincide everywhere. Protein bands in PMCs are weaker than in MCs. Apparently, PMCs <0.2 microns contain less protein.

The significant difference between MCs and PMSs in the intensity of a wide structure-less band in the range from 4000 cm^{-1} to 2000 cm^{-1} is due to the presence of more bound water in the MC, as well as greater light scattering (therefore, when calculating the intensities of protein bands, the contribution of this structure-less band was subtracted).

Judging by the intensities of the protein and nucleotide IR bands, the protein/DNA ratio in PMCs is lower than in MCs, i.e. germinal PMCs are poorer in protein.

Based on the data obtained, we can propose an improved scheme (Fig. 7.18) of the MC life cycle in specialized animal cells. Small germinal PMC with a size of about 0.1 μm arise due to DNA leaving the cytoplasm from post-MC or from degraded MC. For the synthesis of PMC proteins, the nuclear DNA is activated (it is not yet known what exactly is the signal for the activation of the required nuclear DNA genes). Damaged, oxidized, or mutated DNA molecule from MC, unable to form germinal PMC, is likely to undergo cleavage in cellular cytoplasm by DNAases and nucleases. In the course of the cell life cycle, PMC emerges and grows (0.1–0.45 microns), followed by growth and development into adult MC, then MC aging to post-MC takes place with accumulation of lipofuscin, up to the formation of lipofuscin granules. If proteases and lysosomes are sufficiently active, then lysis of the old MC occurs (without the formation of large quantities of lipofuscin) with the release of DNA from them into cytoplasm.

The obtained results for PMCs <0.22 μm are in good agreement with the results for PMCs <0.45 μm [50] and confirm the thesis that MCs in specialized cells arise mainly not by division or budding, but from germinal PMCs – particles of size less than 0.2 microns containing DNA and a number of proteins.

Judging by the data of optical microscopy and filtration, the number of germinal PMCs <0.2 microns in the liver cells does not exceed 10% of the number of matured MCs. This is consistent with the data of optical dark-field microscopy of cells [45] and suspensions of different MC fractions [56].

Conclusion Germinal PMCs generally contain the same proteins as MCs. However, the content of flavoproteins and especially cytochrome oxidase is reduced in them. Germinal PMCs, unlike MCs, almost do not contain the "aging pigment" lipofuscin. They have DNA in matrix, moreover, the ratio of DNA/protein in them is higher than in the MCs, i.e. they are depleted in protein. The obtained results confirm that MCs in specialized animal cells can arise from germinal PMCs – particles less than 0.2 microns in size containing DNA. It is assumed that the DNA entering the cytoplasm during the degradation of old MCs serves as a seed for the formation of PMCs (in cooperation with nuclear DNA).

The results were obtained in collaboration with A.V. Chaplygina and V.I. Kovalev. The paper is accepted in BBR, 2019.

7.4 Fluorimetric Comparison of Proto-mitochondria and Mitochondria

MCs are heterogeneous in size, morphology [57, 58] and density [59, 60]. They participate in the ontogenesis of the cell and aging [61]. MCs of different sizes are organelles at different stages of the life cycle [57, 60, 62, 63]. In any specialized mammalian cells, three populations are simultaneously present: "young" – with a diameter of less than 1 micron, "mature" – about 1 micron, and "old" – larger than 1–2 micron [57].

Young MCs in specialized animal cells usually arise not by division, but from the smallest germinal PMCs [57, 64], named so [64] by analogy with yeast pro-mitochondria [65, 66].

Pro-mitochondria in yeast are organelles with undeveloped membranes and without some cytochromes [65, 66]. They were first isolated and studied many years ago [66–70]. Like MCs, they contain circular DNA. The activity of ATPase, NADH: ferricyanide reductase and succinate: ferricyanide reductase of yeast pro-mitochondria is several times lower than their MCs. The NADH oxidase, succinate oxidase and cytochrome oxidase activities are not detected at all, which indicates an incomplete or damaged respiratory chain. When yeast is placed under aerobic conditions, their pro-MCs become capable of protein synthesis and transformation into normal MCs [69, 70].

Unlike the well-studied yeast *pro*-mitochondria, animal *proto*-mitochondria have not been studied much. Their isolation in intact form was first carried out in [71]. PMC with a diameter of 0.45 microns possess both membranes – external and internal [71, 72]. In cells, young PMC gradually has increasing in size to 1 micron, become the mature MC [73–77]. The biogenesis of MC membranes is known in detail [78, 79]. Some subunits of NADH dehydrogenase, cytochrome oxidase and ATPase are encoded by mitochondrial DNA and synthesized on MC ribosomes [80, 81]. At least 85% of the MC proteins are encoded by nuclear DNA [78–82].

Fixed-diameter PMCs were first obtained from mammalian heart and liver cells using high-speed centrifugation followed by filtration of the diluted total mitochondrial fraction through inert millipore filters with a small pore diameter [64, 71]. A biochemical study of two fractions of PMCs was carried out: with a diameter of less than 0.45 microns and less than 0.22 microns [71] in comparison with mature MCs.

It was interesting to compare a number of PMC properties with not very mature MCs, smallest than mature MCs.

A light mitochondrial fraction, consisted of PMCs and immature MCs, was obtained from the rat liver. The medium contained 20 mM Tris-HCl and 300 mM sucrose (pH 7.6). Isolation was carried out at 2 °C using the standard method [83] with modifications [84], with the difference that the initial total suspension was centrifuged at 1000 g for 10 min, and the obtained supernatant was centrifuged at 15,300 g for 20 min, as a result which precipitated the lung (light) mitochondria (LMC) and PMC. Not less than 50% of all particles of the total fraction were PMCs, having a size of up to 0.45 microns. The respiratory control of the light fraction reached 2.5 (on succinate and ADP). Separate fixed-size PMC fractions were obtained by filtration of the light fraction (diluted in the isolation medium to a protein concentration of 1–3 mg/ml) through Millex HV millipore filters with pore diameters of 0.45, 0.22, and 0.1 μm or through similar Acrodisc HT Tuffryn filters.

Protein concentration was determined by the biuret method and the UV method [85]. In the light fraction, the protein concentration inicially was 30 mg/ml. Particle size was controlled by correlation microscopy [86], as well as transmitted light microscopy when stained with Janus green.

NADH oxidase, NADH: ferricyanide reductase and NADH: tetrazolium reductase activity of the PMC and LMC fractions were measured after freezing and thawing, leading to the appearance of membrane permeability for NADH. The oxidation rate of NADH was detected photometrically by the decrease in its

Table 7.7 Fluorescence parameters of PMCs and LMCs

Parameters	PMC-0.22	PMC-0.45	LMC
Tryptophan intensity (%)	100	99	93
Maximum (nm)	338	338	337
Half-width (nm)	62	60	58
Life-time (ns)	3.2	3.4	3.6
Polarization degree	0.21	0.21	0.22
Flavin intensity (%)	91	92	100
Maximum (nm)	520	522	524
Half-width (nm)	63	64	65
Life-time (ns)	6.0	6.2	6.5
Polarization degree	0.36	0.37	0.38

Note: the degree of polarization P of tryptophan fluorescence was measured at 338 nm (excitation 295 nm), and flavin – at excitation 450 nm and emission 525 nm

absorption at 340 nm or by the appearance of the absorption of formazan at 530 nm during the reduction of pNTV [57], as well as fluorimetrically, by reducing the fluorescence of NADH (upon excitation of 340 nm and radiation of 440 nm) [87]. Succinate: tetrazolium reductase activity in the same samples was determined by the formation of formazan from pNTV [57].

Fluorescence of tryptophan residues of the PMC and LMC proteins, flavins, and added NADH were recorded on SLM-4800 and Perkin-Elmer MF44 in 1 cm quartz cuvettes at 20 °C. When measuring the lifetime (τ) and the degree of polarization of fluorescence (P) in a number of experiments where the intensity was low, mirror cuvettes were used [88]. Absorption spectra were recorded on M-40 in a special compartment.

The relative DNA/protein content was determined by the optical density at 260/280 nm after the destruction of the particles with SDS detergent, by the fluorescence intensity of tryptophan residues of proteins, and also ethidium bromide incorporated into DNA [88].

A number of fluorescence properties of proteins and flavo-proteins of the PMC and LMC fractions were compared (Table 7.7). For this purpose, their fluorescence was measured (in the range of 315–410 nm, upon excitation at the "red" edge of the tryptophan absorption band of 300 nm) and their flavin fluorescence (in the range of 500–580 nm, upon excitation at the maximum of the flavin absorption band at 450 nm).

Tryptophan Fluorescence The intensity of tryptophan fluorescence of PMCs and LMCs was almost the same (Table 7.7; with aligned optical densities of the samples in the range 280–300 nm). It should be noted that the concentration of the LMC protein, measured by the biuret reaction or by the UV method in the presence of SDS, was 25–35% higher than PMCs. This is due to the larger size of each LMC. That is why LMCs scatters light stronger and is therefore less involved in absorption. The half-width of the spectrum, the lifetime, and the degree of polarization in all

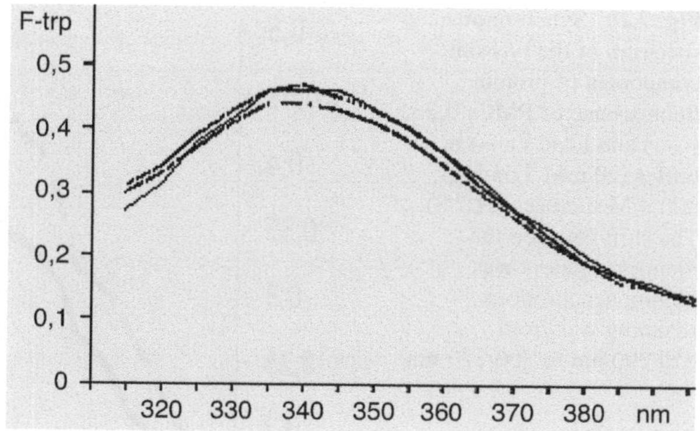

Fig. 7.19 Spectrum of tryptophan fluorescence of PMCs 0.22 (- - -), PMCs 0.45 (...) and LMCs (—) in SDS. Excitation was 300 nm

samples also differed slightly. This means that the composition of the proteins and conformational mobility of components of PMCs and LMCs are very similar.

The intensity of tryptophan fluorescence of proteins after treatment of PMCs and LMCs by SDS completely destroying all particles (the total disappearance of weak light scattering was observed) also was almost the same (Fig. 7.19).

Small differences are visible only in the short wavelength region of the spectrum, where significant contribution belongs to tyrosine residues. It can be assumed that the difference is due to different ionization of tyrosines or enhance in fluorescence quantum yields of tyrosine complexes capable of absorbing UV even at 290–300 nm [89]. This is evidenced by the difference in the intensity of a single tyrosine component detected by synchronous scanning of monochromators (Fig. 7.20; the method of synchronous tyrosine fluorimetry is described in [88]); such tyrosine emission of PMCs is more intense than LMCs.

Flavin Fluorescence The fluorescence of flavins was slightly less in the PMC fraction than in the LMC. This means that either PMCs contain less flavoproteins, or flavins in them are more strongly quenched by iron-sulfur clusters [86]. The half-width of the spectrum, the lifetime, and the polarization degree of flavin fluorescence in the samples differ little (Table 7.7 and Fig. 7.21).

The lifetime and polarization degree of the PMC fraction are slightly less than of LMC fraction. According to the Levshin – Perrin equation [88], substituting the lifetime and polarization degree and assuming the flavin volume to be 100 Å^3, the viscosity of the microenvironment can be calculated. It is very high – 15 poise (more viscosity of glycerol). This means that in both PMCs and in LMCs, flavin molecules have minimal rotational mobility, and the viscosity of the microenvironment of flavins in PMC is slightly lower than in LMC.

Upon the destruction of PMCs and LMCs by SDS, the difference in flavin fluorescence increased sharply: in LMCs, the intensity is 30% higher (Fig. 7.21). This means that PMCs do contain less flavoprotein than LMCs. Since the main contribution to the flavin fluorescence of MCs is made by the flavin of NADH dehydrogenase [64, 90], it can be assumed that upon maturation of PMC to MC, the

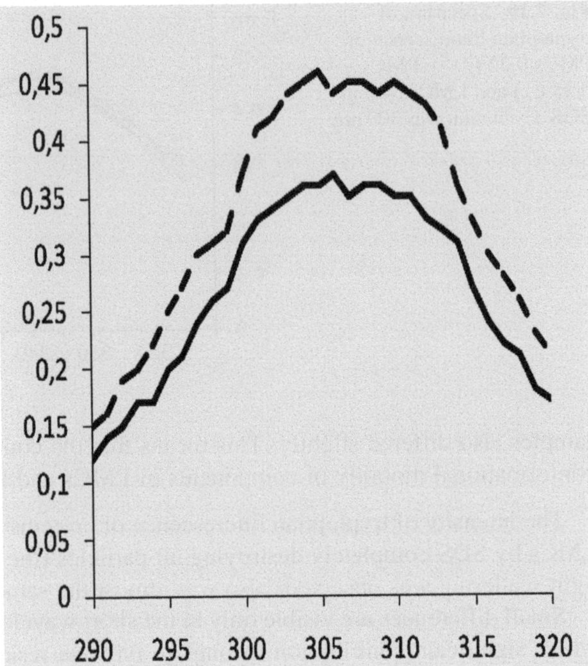

Fig. 7.20 Synchronous spectrum of the tyrosine component of protein fluorescence of PMCs 0.22 (- - -) and LMCs (—) in buffer (20 mM Tris-HCl, 300 mM sucrose, pH 7.8). The shift between the monochromators was 20 nm; synchronous scanning was from 270/290 nm to 300/320 nm

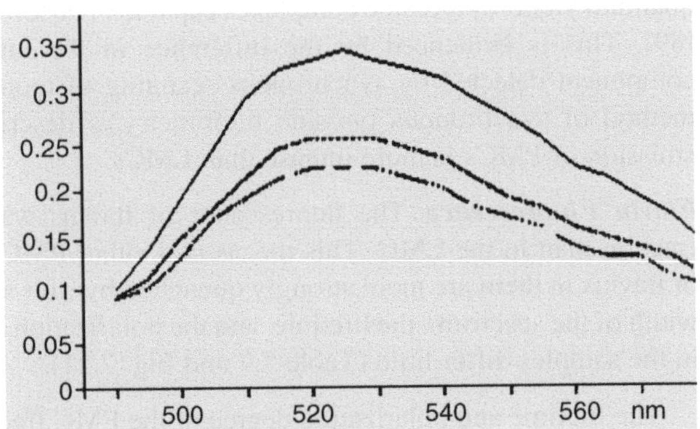

Fig. 7.21 Fluorescence of flavins of PMCs 0.22 (- - -), PMCs 0.45 (...) and LMCs (—) in SDS. The wavelength of the exciting light was 450 nm

amount of NADH dehydrogenase increases. At the same time, it was previously discovered [71] that the respiratory activity of PMCs and the activity of their NADH dehydrogenase are even slightly higher than that of mature MCs. Apparently, in growing organelles, a decrease in the activity of this enzyme is compensated by an increase in its amount. The most important cause of the loss of NADH oxidase respiratory activity is the spontaneous "dumping" of flavin (in the form of FMN or riboflavin) [90, 91].

NADH Oxidase Activity Intact PMCs, as well as intact LMCs, oxidize added NADH too poorly due to low permeability of inner membrane. After freezing, the

Table 7.8 The rate (μM/min/mg protein) of the oxidation of NADH and succinate in the PMC and LMC fractions

Additions:	PMC < 0.1	PMC < 0.22	PMC < 0.45	LMC
NADH	0.06	0.09	0.16	0.1
NADH + NAD	0.03	0.04	0.09	0.06
NADH + rotenone	0.01	0.02	0.04	0.02
NADH + ferricyanide	0.8	0.49	0.37	0.47
NADH + ferricyanide + NAD	0.36	0.28	0.27	0.3
NADH + pNTV	0.5	0.44	0.38	0.43
NADH + pNTV + NAD	0.25	0.26	0.27	0.26
NAD**P**H	0	0	0	0
Succinate + pNTV	0.02	0.07	0.14	0.08
Succinate + pNTV + malonate	0.01	0.03	0.07	0.04

Note. The first five reactions were measured fluorimetrically, the rest – photometrically. Reaction measurement time was 5 min

permeability increases sharply, and the ability to oxidize the added NADH arises quickly enough [84]. At the same time, however, sensitivity to rotenone and other respiratory chain inhibitors is markedly reduced. The rotenone-insensitive oxidation reaction of NADH at the level of complex I is an electronic shunt (bypassing ubiquinone and cytochromes of the respiratory chain) [84] immediately to oxygen, which is reduced to superoxide.

In all fractions of PMCs, the NADH oxidation was observed. Moreover, its speed in PMCs 0.45 was even higher than in LMCs (Table 7.8). The reaction proceeds precisely at the level of NADH dehydrogenase of the respiratory chain, since an analogue like NAD**P**H, unlike NADH, was not oxidized (Table 7.8).

The oxidation of NADH in MCs is known to be suppressed by the reaction product – NAD [84]. The oxidation of NADH in PMC fractions was markedly suppressed by the addition of an excess amount of NAD (Table 7.8).

NADH: Ferricyanide Reductase Activity PMCs are more active than mature MCs in catalysis of the NADH: ferricyanide reductase reaction [71]. NADH oxidation, induced by artificial electron acceptors like ferricyanide, occur an order of magnitude faster than rotenone-sensitive oxidation in the respiratory chain [84, 91]. The initial rate of this reaction for the smallest PMC – 0.1 μm was 0.8 μM/min/mg protein. A quick reaction with ferricyanide is an electronic shunt bypassing flavin and ubiquinone [84]. Rotenone had practically no effect on the rate of this reaction in either LMCs or PMCs, as one would expect.

NADH oxidation can be observed not only photometrically, but also by the disappearance of its fluorescence. The smallest germinal PMCs <0.1 μm have much higher NADH: ferricyanide reductase activity than LMCs (Fig. 7.22). Adding excess NAD slows down the process (Table 7.8).

NADH: pNTV Reductase Activity When using pNTV, the oxidation rate of NADH was lower than in the case of ferricyanide in all samples. This reaction is a shunt

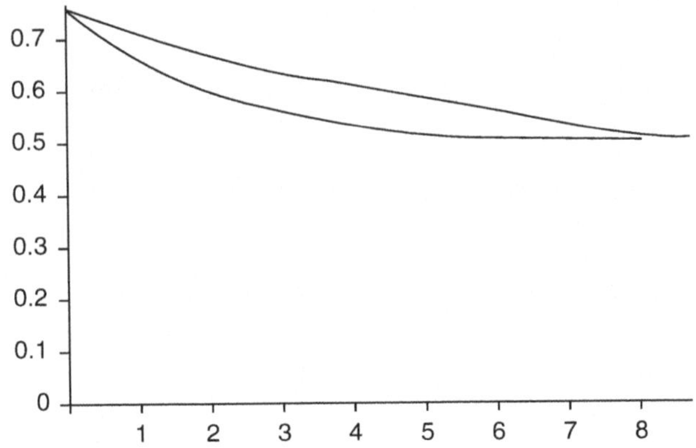

Fig. 7.22 Kinetics of NADH: ferricyanide reductase reaction, detecting by NADH fluorescence for 8 min with LMCs (-) and PMCs 0.1 (...)

Table 7.9 The fluorescence intensity (F) of the probes in the fractions of proto-mitochondria and light mitochondria

Dye	λex (nm)	λem (nm)	*F* PMC 0.1	*F* PMC 0.45	*F* LMC
ANS	370	450	56 ± 4	59 ± 4	51 ± 4
Mer-540	540	580	150 ± 5	170 ± 5	190 ± 5
Pyrene monomers	338	393	100	100	100
Pyrene excimers	338	480	33	33	33
Tetracycline	390	535	44	43	38
Ethidium bromide	470	610	13	14	15

bypassing flavin and ubiquinone [64, 87]). The reaction rate was decreased by NAD (Table 7.8), but not by rotenon.

Succinate: pNTV Reductase Activity The pNTV reduction rate in the case of succinate instead of NADH is rather low. When a malonate (succinate dehydrogenase competitive inhibitor) with succinate was added, the reaction speed sharply decreased. This suggests that the reaction proceeds precisely on succinate dehydrogenase (no any back transfer of electrons to NADH dehydrogenase). With increasing size of the PMCs, the reaction rate increases significantly. Apparently, this is due to an increase in the amount of succinate dehydrogenase in more mature organelles.

DNA Content The relative DNA content (with respect to the protein) in LMCs and in PMCs is quite close. So, the optical density at 260 nm, where the main contribution to the absorption of MCs belongs to DNA (as well as some RNA and ubiquinone), was almost the same for all fractions. The difference between the maximum and minimum fluorescence intensities of DNA-specific ethidium bromide was small (Table 7.9). This means that the growth of PMCs in cells is accompanied not only by an increase in protein mass, but also in parallel by an increase in the number of DNA copies. This is consistent with similar results [74–76] obtained previously by other authors using different methods at other objects.

Fluorescent Probes Four membrane-tropic fluorescence probes were used to characterize a number of physicochemical properties of PMC membranes: ANS, merocyanine-540, tetracycline, and pyrene. ANS primarily responds to charge, hydrophobic pockets, and protein conformation; merocyanine – to the membrane potential and phase state of lipids, tetracycline – to calcium and positive charges, pyrene – to the viscosity of the lipid phase and membrane molecular oxygen [88, 89].

The fluorescence spectra of all the probes used in all fractions of PMCs in form did not practically differ from those in the LMC. This indicates the similarity of their membranes in polarity and charge, hydrophobic pockets and protein conformation, membrane potential, phase state and viscosity of lipids. At the same time, the fluorescence intensity of the probes was not the same everywhere (Table 7.9). Therefore, PMC membranes are somewhat different from LMC membranes in the number of charges or hydrophobic pockets and protein conformations. Interestingly, the longer the suspensions were stored, the less the differences between MCs and PMCs detected by fluorescent probes became [92].

The difference between the intensities of merocyanine-540 in the 0.1-μm PMCs and in LMCs reached 20% (Table 7.9). Apparently, the membrane potential and/or phase state of lipids of the PMC membranes differ from those of LMCs. The result depended on the age of the suspensions: the longer the suspensions were stored, the smaller the differences [92].

The fluorescence intensity of pyrene (monomers of pyrene and its excimers and dimers) for all fractions was exactly the same (Table 7.9). The monomer pyrene lifetime $\tau = 105$ ns also coincided. This indicates the same viscosity of lipids and the same oxygen content in the membranes of PMCs and LMCs. The intensity of monomeric emission of pyrene in all fractions was 1.1–1.4 times lower than in water, i.e. in these membranes, pyrene is quenched by iron-sulfur clusters, hem proteins, and oxygen (in other types of membranes, for example, in the sarcoplasmic reticulum, pyrene fluoresces more strongly than in water [88]).

Conclusion Using centrifugation and filtration through millipore filters with pore sizes from 0.1 to 0.45 μm, intracellular germinal organelles, PMCs, which are precursors of LMCs in specialized animal cells, were isolated and characterized from rat liver cells. PMCs do not differ significantly from LMCs in the general properties of proteins, lipids, DNA, surface charge, and other parameters. At the same time, the activity of some respiratory chain enzymes (NADH dehydrogenase, succinate dehydrogenase) is even higher in PMCs than in LMCs. The results were published [93].

7.5 Photometry and Fluorimetry of Proto-mitochondria from Liver of Young and Adult Rats

In specialized cells of different organs (muscles, heart, brain, liver, etc.), MCs usually appear not by division, but by growth from germinal organelles of 0.1 microns in size. The "seed" of these organelles is 16 kbs DNA released after damage or lysis of old post-MCs. Particles with a diameter of 0.2–0.45 μm are PMCs – small young organelles with not too much respiratory activity but high activity of NADH: ferricyanide reductase, NADH: tetrazolium reductase, succinate: ferricyanide reductase and succinate: tetrazolium reductase. PMCs 0.45 microns in size possess both membranes – external and internal.

The properties of PMCs of animal cells are very different from the properties of yeast proMCs [94] – aberrant organelles with a single membrane, lacking certain cytochromes and with low NADH: ferricyanide reductase activity and succinate: ferricyanide reductase activity, as well as with zero activity of NADH oxidase, succinate oxidase and cytochrome oxidase.

Fixed-diameter PMCs were first obtained from heart cells and then from liver cells by centrifugation followed by filtration through calibrated millipore filters [95]. PMCs ranging in size from 0.22 to 0.45 microns are similar to MCs in total protein composition and protein/DNA ratio.

PMCs in specialized mammalian cells coexist simultaneously with MCs. The quantitative ratio of the PMC and MC populations depends on the type of cells and their age. In young and adult animals, the number of PMCs and some of their properties can vary markedly.

It was interesting to study the possibility of mass transformation of PMCs to MCs in adult rats, as well as to find out the similarities and differences in PMCs in young and adult rats by the properties of membrane proteins, in particular, cytochromes.

The mitochondrial fraction from the liver of male young (1 month old) and very adult (9 month old) Wistar rats was isolated in a medium containing 20 mM Tris-HCl and 300 mM sucrose (pH 7.7), at 2 °C. First, the total suspension was centrifuged at 600 g for 7 min. The supernatant was centrifuged for 15 min at 2800 g, as a result of which it precipitated into mature heavy MC, which were then suspended in the same medium (9 ml per 1.5 ml of sediment), packaged in 1 ml Eppendorf tubes (in the cold) and used for experiments. The secondary supernatant was centrifuged at 14,000 g for 17 min, whereby light mitochondrial organelles precipitated. The resulting pellet of light MC (LMC) was suspended in the same medium to the concentration of 2–3 mg protein/ml, after which a part of the LMC was packaged in tubes, and another one was passed through MILLEX-GV millipore filters with a pore diameter of 0.45 or 0.22 μm to obtain a PMC with a diameter up to 0.45 microns (PMC-0.45) and up to 0.22 microns (PMC-0.22). All fractions were aligned by protein content. Particle size was controlled by correlation microscopy, dark-field microscopy, and transmitted light microscopy when stained with Janus green.

Table 7.10 The content of organelles of a certain diameter in the light mitochondrial fraction of young and adult rats

Diameter of organelles (μm)	Content (%) of organelles in fractions	
	LMC-Y	LMC-A
<0.22	43 ± 3	36 ± 3
0.22–0.45	22 ± 3	39 ± 3
>0.45	35 ± 3	25 ± 3

Absorption spectra in the UV and visible region were recorded on M-40 spectrophotometer. Moreover, in a number of experiments (to eliminate light scattering and to correctly compare the content of components by band intensities), the samples were treated with SDS, which has no absorption and fluorescence neither in the UV nor in the visible region.

The fluorescence spectra of ANS and tryptophan residues of PMC, LMC and MC proteins were recorded on SLM-4800 in 1 cm quartz cuvettes at 20°C. In a number of experiments where the intensity was low, in particular, when measuring the polarization degree (P), mirror cuvettes were used [96].

Filtration The number of mitochondrial organelles of a certain size is easily found by filtering diluted suspensions through millipore filters of a fixed diameter (with control of the initial and final optical density at 280 nm, after the destruction of the particles by SDS). The ratio of organelles of different sizes in the LMC of the liver of young and adult rats was not the same (Table 7.10). In young rats, the smallest proportion is made up of particles with diameters from 0.22 to 0.45 microns, and in adults, over 0.45 microns. Organelles up to 0.45 microns in size are mostly PMC.

It should be emphasized that PMC, isolated from the light fraction, are no any fragments or sub-mitochondrial particles. PMCs can be easily seen in living liver cells (and in the cells of all other organs) in a dark-field mode, as well as in transmitted light when stained with formazan, formed during the succinate: tetrazolium-reductase reaction. Moreover, the total number of different PMCs before and after filtering does not change.

Absorption Spectra In the absorption spectra of MC suspensions of young and adult rats, cytochrome peaks with a maximum at the Soret band near 410 nm are clearly visible, and these peaks are less pronounced in the corresponding PMC fractions. Moreover, in young rats, all fractions had spectra slightly different in form from the spectra of adults. Since MCs have strong light scattering, which hinders the determination of peak values (spectra are "smeared" even when suspensions are placed in a special compartment of spectrophotometer for turbid samples), it is difficult to quantify cytochrome content. Light scattering of the MC fraction from adult individuals is bugger, since their MCs is a bit larger than from young rats. Therefore, the absorption spectra of the fractions were recorded after treatment by SDS (1%), which destroys the organelles and almost completely eliminates light scattering.

Table 7.11 The cytochrome content in relation to the total amount of proteins and DNA, measured by the ratio of optical densities at 405 and 265 nm in proto-mitochondria and mitochondria of young and adult rats after treatment of these fractions by SDS (1%)

	PMC-0.22-Y	PMC-0.22-A	PMC-0.45-Y	PMC-0.45-A	LMC-Y	LMC-A	MC-Y	MC-A
*D*405/*D*265	0.1	0.11	0.12	0.13	0.14	0.14	0.23	0.18

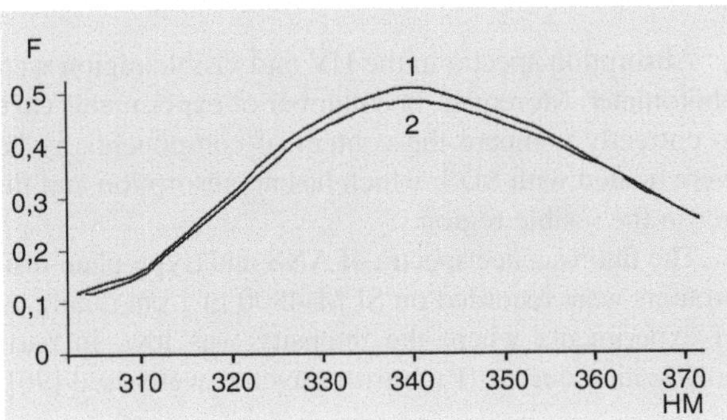

Fig. 7.23 Spectra of tryptophan fluorescence of PMC-0.22 from adult (1) and young (2) rats. Excitation was 290 nm

Table 7.12 The ratio of the intensities of protein fluorescence of mitochondrial fractions in the short and long wavelength parts of the spectrum (at 310 and 370 nm, with respect to the intensity at maximum, at 338 nm). Excitation was 290 nm

	PMC-0.22-Y	PMC-0.22-A	PMC-0.45-Y	PMC-0.45-A	LMC-Y	LMC-A
*F*310/*F*338	0.39	0.4	0.41	0.42	0.41	0.4
*F*370/*F*338	0.52	0.48	0.52	0.47	0.53	0.53

Table 7.11 shows data on the spectrophotometrically determined cytochrome content (relative to the total amount of proteins and DNA) in solutions obtained by treatment of mitochondrial fractions by 1% SDS. It can be seen that the number of cytochromes in PMC is noticeably lower than in MC, and their minimum content is in small PMC-0.22. This fact reflects a gradual increase in the content of cytochromes in PMC as they intracellularly mature to MC. Interestingly, in the MC of young rats, there are almost a quarter more cytochromes than in adults, i.e. as animals grow older, the content of cytochromes, on the contrary, decreases. This may be due either to their loss or to deterioration in synthesis (Table 7.11).

Tryptophan Fluorescence The intensity of tryptophan fluorescence of PMC and MC was similar, which indicates the same or almost the same amount of tryptophan residues. Figure 7.23 shows the spectra of tryptophan fluorescence of PMC-0.22 in adult and young rats. The half-width of the spectrum was slightly different, moreover, due to the long-wavelength component (Table 7.12).

Although the ratio between the intensities at 310 nm is almost the same for all samples (under normalization conditions at 338 nm, at the maximum of the

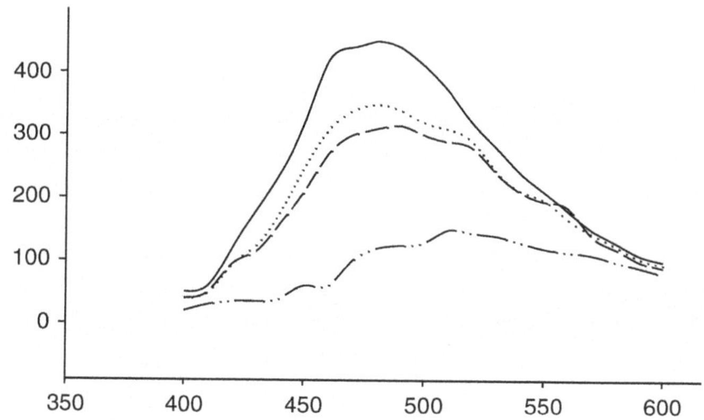

Fig. 7.24 Fluorescence spectra of ANS (10^{-7} M) in the fractions PMC-0.22 (-), PMC-0,45 (...) and MC (- - -) of adult rats; and the ANS spectrum in the buffer (- .. - ..). Excitation was 370 nm. The amount of protein in the fractions was 0.3 mg/ml

Table 7.13 ANS fluorescence intensity (F) and polarization (P) of mitochondrial fractions (suspensions in buffer)

	PMC-0.22-Y	PMC-0.22-A	PMC-0.45-Y	PMC-0.45-A	LMC-Y	LMC-A	MC-Y	MC-A
F	1.2	1.35	0.72	1.05	0.71	1.02	0.71	0.98
P	0.3	0.28	0.3	0.28	0.29	0.29	0.31	0.32

Note: ANS concentration – 8 μM; excitation – 370 nm, emission – 460 nm

spectrum), it differs at 370 nm (Table 7.12). This means that there are variations of the long-wave component. The long-wavelength band is characteristic of proteins, where tryptophan residues come into contact with the aqueous phase [96]. Apparently, PMC (0.45 and 0.22 microns) of adult individuals contains ~10% more non-folded (denatured) proteins than juveniles.

Fluorescence of ANS PMCs and MCs markedly differ in the intensity of fluorescence of hydrophobic probes that bind to membrane proteins [96], in particular, in the case of ANS. In MCs, ANS fluoresces mainly from proteins, and not from lipid phase [96].

Figure 7.24 shows the ANS fluorescence spectra in the PMC-0.22, PMC-0.45, and MC fractions of adult rats. No spectral shifts were detected between the fractions. It indicates a close polarity of the microenvironment for the probe. But the intensity varies markedly. It was maximal for PMC-0.22 and minimal for MC. This can be caused by a different number of ANS binding sites in membrane proteins. In mature MC, the proportion of proteins belonging not to the membrane, but to the matrix, to which the ANS penetrates poorly, is increased.

Table 7.13 shows the data on the ANS fluorescence intensity in fractions. Differences in the ANS fluorescence intensity in paired fractions from the liver of young and adult rats reach 30–40%. Moreover, the minimal difference (11%)

between young and adult individuals is found only in the smallest (immature) organelles – PMCs-0.22. The main reason for the differences is the different number of ANS binding sites in membrane proteins. In the process of maturation of mitochondrial organelles, a decrease in the proportion of membrane proteins probably occurs. This is observed in both young and adult individuals. Moreover, the amount of such proteins in young individuals grow up to adults also decreases.

The differences in the ANS fluorescence intensity is caused by just a different number of binding sites, and not by a change in physicochemical properties. It was confirmed by measuring the fluorescence polarization degree of ANS. No strong differences between the samples are detected (Table 7.13). Since, as already mentioned, ANS in MCs strongly fluoresces mainly from the protein and not from the lipid phase, that clear from high polarization – about 0.3. A high polarization degree of ANS occurs in proteins due to their very big microviscosity, which impedes probe rotation [96]. The almost identical magnitude of the ANS polarization in different fractions indicates the identity or, at least, the proximity of their membrane physicochemical properties – microviscosity, polarity and surface charge.

The data obtained allow us to conclude that in the cells of young animals there is a massive transformation of PMCs into MCs, similar to that observed in adult rats. The transformation is accompanied, in particular, by an increase in the number of cytochromes. As animals grow older, the number of cytochromes in MCs, on the contrary, decreases. The results were published [97].

References

References for Section 7.1

1. Luck DJ (1963) PNAS 49(2):233–240
2. Luck DJ (1965) J Cell Biol 24(3):461–470
3. Anderson E, Condon W, Sharp D (1970) J Morphol 130(1):67–91
4. Ozernyuk ND (1978) Growth and reproduction of mitochondria. Science, Moscow
5. Stephens RJ, Bils RF (1967) J Ultrastruct Res 18(3):456–474
6. Belyakovich AG (1990) Study of mitochondria and Bacteria using Tetrazolium salt pNTV. ONTI, Pushchino
7. Sharova IV, Vekshin NL (2000) Mitochondria, cells and reactive oxygen species, Pushchino, pp 163–166
8. Sukharev VI, Vekshin NL (2000) Bioorg Chem 26(4):723–726
9. Lomtev AS, Sharova IV, Vekshin NL (2001) Dokl Acad Sci 376(1):114–116
10. Panov AV (2015) Practical mitochondriology. Novosibirsk
11. Meshkova MP, Severin SE (eds) (1969) Workshop on Biochemistry. Moscow State University, Moscow, pp 354–357
12. Vekshin NL (2002) Photonics of biopolymers. Springer
13. Sharova IV, Vekshin NL (2004) Biophysics 49(5):814–821
14. Vinogradov AD et al (1999) Biochemistry 64:174–193
15. Walker JE (1992) Q Rev Biophys 25:253–324
16. Lehninger A (1966) Mitochondria. Mir, Moscow

17. Zinina AN, Vekshin NL (2008) Biological motility, vol 1, Pushchino, 190–194
18. Shishmakov DA, Anisimov RL, Vekshin NL (2006) Biol Membr 21(5):374–380

References for Section 7.2

19. Friedman JR, Nunnari J (2014) Nature 505:335–343
20. Bereiter-Hahn J (1995) Int Rev Sci 30:99–110
21. De Vos KJ (2005) Curr Biol 15:678–683
22. Mishra NC, Kumar S (2005) Indian J Exp Biol 43:25–34
23. Granville DJ, Gottlieb RA (2002) Sci World J 2:1569–1578
24. Belyakovich AG (1990) Study of mitochondria and bacteria with tetrazolium salt pNTV, Pushchino
25. Shishmakov DA, Anisimov RL, Vekshin NL (2004) Biol Memb 21(5):389–395
26. Zinina AN, Vekshin NL (2008) Biol Membr 25(6):480–487
27. Vekshin NL (2010) Biol Membr 27(5):424–429
28. Vekshin NL (2008) Fluorescence spectroscopy of biopolymers. Photon, Pushchino
29. Howell KA, Millar AH, Whelan J (2006) Plant Mol Biol 60:201–223
30. Vekshin NL (1988) Biol Sci 4:107–111
31. Birnboim H (1979) J Doly Nucl Acids Res 24(6):1513–1523
32. Laemmli UK (1970) Nature 227:680–685
33. Osterman LA (1981) Methods for the study of proteins and nucleic acids. Nauka, Moscow
34. Kosterina EA (2013) Bull Siberia Med 12(3):5–17
35. Forner F, Foster LJ, Campanaro S, Valle G, Mann M (2006) Mol Cell Proteomics 5:608–619
36. Foster LJ (2006) Cell 125:187–199
37. Lanni A, Moreno M, Lombardi A, Goglia F (1996) Int J Biochem Cell Biol 28:337–343
38. Luzikov VN (1980) Regulation of mitochondrial formation. Science, Moscow
39. Begunova EA, Vekshin NL (2015) Biophysics 60(6):1109–1117

References for Section 7.3

40. Friedman JR, Nunnari J (2014) Nature 505:335
41. Bereiter-Hahn J, Voth M (1994) Miscosc Res Techn 27:198
42. De Vos KJ, Allan VJ, Grierson AJ, Sheetz MP (2005) Curr Biol 15(7):678
43. Mishra NC, Kumar S (2005) Indian J Exp Biol 43:25
44. Granville DJ, Gottlieb RA (2002) Sci World J 2:1569
45. Belyakovich AG (1990) Study of mitochondria and bacteria using tetrazolium salt pNTV. ONTI, Pushchino
46. Shishmakov DA, Anisimov RL, Vekshin NL (2004) Biol Membr 1(5):389
47. Zinina AN, Vekshin NL (2008) Biol Memb 25(6):480
48. Vekshin NL (2010) Biol Membr 27(5):424
49. Vekshin NL (2008) Fluorescence spectroscopy of biopolymers. Photon, Pushchino
50. Begunova EA, Vekshin NL (2015) Biophysics 60(6):1109
51. Frolova MS, Surin AM, Braslavsky AV, Vekshin NL (2015) Biophysics 60(5):1125
52. Krekhova KE, Vekshin NL (2018) Biophysics 63(2):265–269
53. Vekshin NL (2002) Photonics of biopolymers. Springer
54. Lehninger A (1976) Biochemistry. Mir, Moscow
55. Vekshin NL (2011) Biophysics of DNA-antibiotic complexes. Nova Science Publishers, NY
56. Kurdyukov DN, Vekshin NL (2016) Biophysics 61(4):736

References for Section 7.4

57. Belyakovich AG (1990) Study of mitochondria and bacteria using tetrazolium salt pNTV. ONTI, Pushchino
58. Logan DC, Leaver CJ (2000) J Exp Bot 51:865–871
59. Lanni A, Moreno M, Lombardi A, Goglia F (1996) Int J Biochem Cell Biol 28:337–343
60. Dai H, Lo YS, Jane WN, Lee LW, Chiang KS (1998) Eur J Cell Biol 75:198–209
61. D'Herde K et al (2000) Cell Death Differ 7:331–337
62. Bereiter-Hahn J (1990) Int Rev Cytol 122:1–63
63. Bereiter-Hahn J, Voth M (1994) Miscosc Res Tech 27:198–219
64. Lomtev AS, Sharova IV, Vekshin NL (2001) Dokl Acad Sci 376(1):114–116
65. English-Russian dictionary of genetic and cytological terms. 1973 Novosibirsk
66. Criddle RS, Schatz G (1969) Biochemistry 8:322–334
67. Paltauf F, Schatz G (1969) Biochemistry 8:335–339
68. Plattner H, Schatz G (1969) Biochemistry 8:339–343
69. Plattner H, Salpeter MM, Saltzgaber J, Schatz G (1970) Proc Natl Acad Sci 66:1252–1259
70. Groot GS, Rouslin W, Schatz G (1972) J Biol Chem 247:1735–1742
71. Shishmakov DA, Anisimov RL, Vekshin NL (2004) Biol Memb 21:389–395
72. Zinina AN, Vekshin NL (2008) In: Biological motility. Pushchino, vol 1. pp 190–194
73. Stephens RJ, Bils RF (1967) J Ultrastruct Res 18(3):456–474
74. Rudin D, Wilkie D (1970) Biogenesis of mitochondria. Mir, Moscow
75. Ozernyuk ND (1978) Growth and reproduction of mitochondria. Science, Mjscow
76. Luzikov VN (1980) Regulation of mitochondrial formation. Science, Moscow
77. Anderson E, Condon W, Sharp D (1970) J Morphol V. 130. N 1, 67–91
78. Tokatlidis K, Schatz G (1999) J Biol Chem 274:35285–35288
79. Schultz B, Chan S (2001) Annu Rev Biophys Struct V 30. pp 23–65
80. Carroll J, Shannon R, Fearnley I, Walker J, Hirst J (2002) J Biol Chem. V 227. pp 50311–50317
81. Bai YD, Attardi G (1998) EMBO J 17:4848–4858
82. Hirst J, Carroll J, Fearnley I, Shannon R, Walker J (2003) BBA V 1604. pp 135–150
83. Meshkova MP, Severin SE (eds) (1969) Workshop on biochemistry. MSU, Moscow, pp 354–357
84. Sharova IV, Vekshin NL (2004) Biophysics 49(5):814–821
85. Vekshin NL (1988) Biol Sci 4:107–111
86. Sukharev VI, Vekshin NL (2000) Bioorg Chem 26(4):723–726
87. Sokolova IB, Vekshin NL (2006) Biol memb 23(3):217–222
88. Vekshin NL (2002) Photonics of biopolymers. Springer
89. Lakovich J (1986) Fundamentals of fluorescence spectroscopy. Mir, Moscow
90. Sokolova IB, Vekshin NL (2008) Biophysics 53(1):73–77
91. Vinogradov AD, Gavrikova EV, Grivennikova VG et al (1999) Biochemistry 64:174–193
92. Sokolova IB, Shishmakov DA, Vekshin NL (2005) In: Reception and intracellular signaling. IBK RAS, Pushchino, pp 281–284
93. Zinina AN, Vekshin NL (2008) Biol Memb 25(6):480–487

References for Section 7.5

94. Rudin D, Wilkie D (1970) Biogenesis of mitochondria. Mir, Moscow
95. Lomtev AS, Sharova IV, Vekshin NL (2001) Dokl Acad Sci 376(1):114–116
96. Vekshin NL (2008) Fluorescence spectroscopy of biopolymers. Photon, Pushchino
97. Vekshin NL (2010) Biol Memb 27(5):424–429

Chapter 8
Degradation of Mitochondria

8.1 Degradation of Mitochondria to Lipofuscin by Heating and UV-Lighting

MCs in specialized cells become age and then undergo destruction by proteases or (sometimes) being digested by lysosomes during mitophagy [1, 2]. If the activity of proteases and lysosomes is insufficient, then MCs can degrade into granules of lipofuscin (LF). Lipofuscin – from Greek. *lipos* – fat and lat. *Fuscus* – dark, brown. The term is very unfortunate, because it turned out that common lipids in many lipofuscin granules are not detected at all. In addition, the term is perceived as a kind of individual substance, although in fact lipofuscin trash can be completely different.

It has long been established [3, 4] that LF granules predominantly are protein nature. LF granules can also include oxidized lipids and other substances [5–7]. For example, in model experiments [8], conjugates of Schiff bases formed upon the condensation of two trans-retinal and phosphotidylethanolamine molecules possessed fluorescence spectra similar to that of intracellular LF granules. In [9], an LF-like fluorescent conjugate was detected during the interaction of lysine and one product of lipid peroxidation. There are different types of LF-like products, depending on the objects from which they are obtained.

In the cells of organs (liver, heart, etc.), aging MCs are the main source of LF granules. This is due, firstly, to that MCs are about 1 microns in size and therefore clearly visible under an optical microscope; secondly, there are a lot of them (500–5000 pieces per cell) and, thirdly, they generate superoxide, which initiates lipid peroxidation, which leads to the formation of granules of LF.

The amount of LF in cells increases with aging, as well as with pathologies, with organ atrophy. The accumulation of LF in tissues gives them a brown color. Especially sharply darken the heart and liver. The main factors that cause the appearance of LF in cells are still not known in detail. One of the important agents may be superoxide [10, 11].

N. Vekshin, *Biophysics of Mitochondria*,
https://doi.org/10.1007/978-3-030-33853-4_8

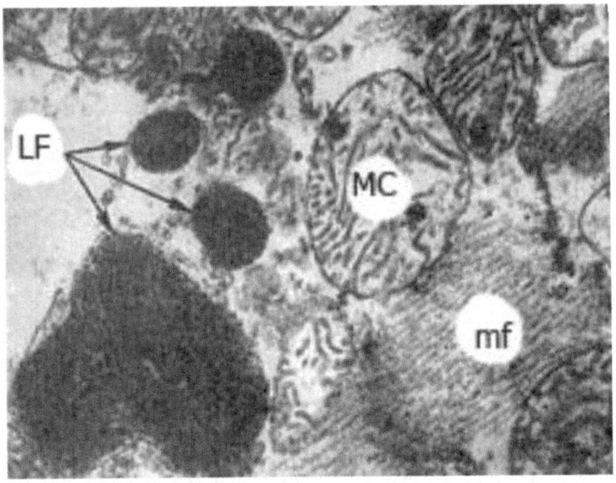

Fig. 8.1 Lipofuscin (LF) in the heart muscle cell, closely associated with mitochondria (MC) and myofibrils (mf). Electron microscopy (x21,000)

The faster MCs age, the more LF is formed in them. The most susceptible to accumulation of LF debris are long-lived post-mitotic (non-dividing) specialized cells: neurons, cardiomyocytes (Fig. 8.1), etc. [11]. Accumulation of LF strongly contributes to the development of neuromuscular degenerative syndromes such as Parkinsonism and brown atrophy of heart [11].

For a long time, the lysosomal mitophagic theory of LF formation was generally accepted [2]. According to this version, damaged and old MCs, falling into active lysosomes, undergo destruction. Although the authors of [12] disrupted the functioning of the ATPase of lysosomes, thereby stopping the hydrolytic activity of lysosomal enzymes, LF in the cells still formed. In recent years, more and more works have appeared that the process of LF formation can proceed without any participation of lysosomes. For example, "artificial" LF was obtained from red blood cells by irradiation with UV light [13]. The formation of LF under UV irradiation of MCs is also shown [14, 15]. Under UV irradiation of proteins via tryptophan and tyrosine residues [16] the denaturation [17] and destroy take place. The rate of photolysis of proteins in solution is usually lower than the rate of photo-induced denaturation [18]. The experiments on UV irradiation of MCs are directly related only to the surface layers of cells: epithelium, cornea, and retina. In [19], LF was initiated in MCs using high (millimolar) concentrations of ascorbic acid and iron (III), thereby starting the well-known Fenton reaction with the formation of fluorescent Schiff bases. Noticeable lipid auto-oxidation and the formation of LF in the MC suspension can proceed even without the addition of iron and ascorbic acid, if forced aeration is carried out—an oxygen purge [20]. These model experiments showed the fundamental possibility of initiating the appearance of LF in MC as a result of lipid peroxydation.

It was interesting to demonstrate the possibility of appearance of LF in the MC suspension *in vitro* in the absence of any added prooxidants and without forced aeration. It is also interesting to compare thermo-induced LF and photo-induced LF.

The MC fraction from rat liver was isolated in a medium containing 20 mM Tris-HCl and 250 mM sucrose (pH 7.7) at 2 ° C [21]. First, the total suspension was centrifuged at 1000 g for 10 min. The supernatant was centrifuged for 15 min at 5000 g, as a result of which mainly mature heavy MCs precipitated, which were suspended in the same medium (9 ml per 1.5 ml of sediment), packed on ice in 1 ml tubes, frozen and then used for experiments.

All samples were aligned with protein concentration using a buffer (100 mM sucrose and 10 mM Tris-phosphate, pH 7.0). MCs were diluted with sterile buffer to a concentration of 0.3 mg/ml (in sterile tubes, under the laminar). Protein concentration was determined by the UV method [22]. The concentration of other proteins – trypsin, lysozyme and bovine serum albumin (BSA) – in model experiments without MCs was also 0.3 mg/ml. The concentration of dissolved nuclear DNA was taken 0.16 mg/ml.

LF formation in the MC suspension was detected by the appearance of a characteristic fluorescence band in the region of 460 nm upon excitation in the region of 360 nm. Moreover, the fluorescence intensity of endogenous pyridine nucleotides under used conditions was negligible. To activate the process of spontaneous formation of LF in MCs, the prolonged moderate heating was applied. For this, a diluted suspension of MCs (0.3 mg/ml), placed in a sterile sealed 15-ml plastic tube, was heated in a thermostat during 24 h at 37 °C or 3 h at 49 °C.

Another way to activate the formation of LF was UV irradiation. The MC suspension in a 1-cm cuvette was irradiated by 450-watt xenon lamp for 1 h, with stirring. In front of all samples a heat filter (a large 2 cm quartz cuvette with water) was used. The temperature of the sample in cuvette at the end of irradiation did not exceed 34 ° C. The temperature in the control was 28 °C.

The fluorescence spectra of the obtained samples were recorded on Perkin-Elmer MPF-44B and Carry ECLIPSE in 1 cm quartz cuvettes at 20 ° C. In a number of experiments where the fluorescence intensity was low, mirror cuvettes were used [23].

The lipid peroxidation level in MCs was determined by thiobarbituric acid (TBA) test for malonic di-aldehyde (MDA) [24]. In a series of experiments, succinate or bicarbonate (300 μM each) were added to the MC suspension (3 mg/ml) in test tubes in a series of experiments and incubated at 37 ° C for 2 h. Then, 2 ml of a 17% solution of TCA and 1 ml of a 3% solution of TAC were added to each sample and then incubated at 60 ° C for 30 min. After cooling to room temperature, the optical density was measured at 532 nm.

One of the important properties of LF is its intense fluorescence in the blue region [1, 5, 13] upon excitation in the near UV region. In the experiments, appearance of intense LF fluorescence after prolonged heating of the MC suspension was observed (Fig. 8.2, right).

A band with a maximum at 360 nm was visible in the excitation spectrum (Fig. 8.2-b, left). A characteristic band of LF with a maximum at 460 nm (upon excitation of 360 nm) appeared in the emission spectrum. It should be noted that a certain amount of LF in the MC suspension was initially, without heating.

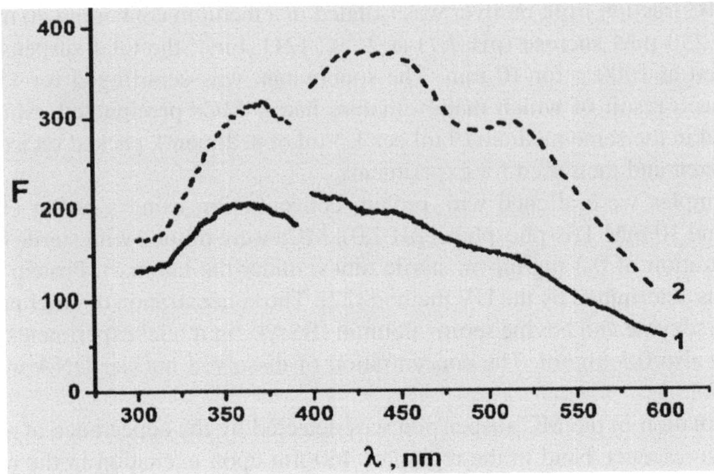

Fig. 8.2 Excitation spectrum (left; emission was 460 nm) and emission spectrum (right; excitation was 360 nm) of MCs (0.3 mg/ml) before incubation (1) and after incubation for 3 h at 49 ° C (2)

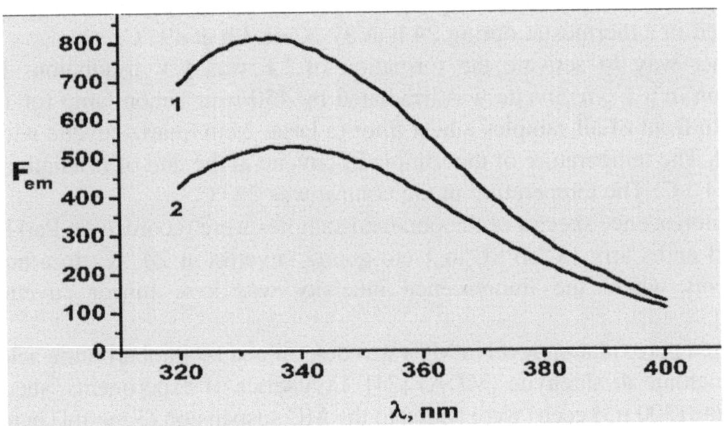

Fig. 8.3 The MC tryptophan fluorescence spectrum (excitation – 286 nm), recorded initially (1) and after 3-h incubation at 49 ° C (2)

To elucidate the origin of the obtained LF, tryptophan fluorescence of the MC suspension was measured before and after heating at 49 °C for 3 h (Fig. 8.3).

After heating, a small long-wavelength shift in the maximum of tryptophan fluorescence of MC proteins occurred from 335 nm to 338 nm. The spectrum broadened slightly (from 60 to 64 nm) and the intensity decreased significantly – by 35%. From this dada we can conclude that there are thermally induced conformational changes in proteins that caused the tryptophan residues to approach to the surface of protein globules or even to aqueous phase. The observed changes are

Table 8.1 The formation of LF after heating and irradiation of various objects in different conditions

Object	LF, %
MCs	100
MCs 2 h at 37 ° C	130
MCs 24 h at 4 ° C	190
MCs 24 h at 37 ° C	270
MCs 3 h at 49 ° C	190
MCs + argon 5 min[a] + 3 h at 49 ° C	160
MCs + irradiation in a glass cuvette 1 h	180
MCs + irradiation for 1 h in a quartz cuvette (hereinafter)	410
MCs + argon 5 min + irradiation 1 h	320
MCs + succinate 30 min + irradiation 1 h[b]	230
BSA + irradiation 1 h	410
Lysozyme + irradiation 1 h	820
Trypsin + irradiation 1 h	240
DNA + irradiation 1 h	110

Note:
[a]After a 5-minute purge of the MC suspension with argon in a sealed cuvette, the amount of oxygen decreased by 50%, after which irradiation was carried out
[b]Pre-incubation with 10 mM succinate for 30 min in an airtight cuvette and subsequent 1-h irradiation

essentially signs of "soft" protein denaturation. This result confirms that LF in MCs is a product of aggregation and cross-linking of denatured proteins.

Aging of isolated MCs in suspension during long-term storage without heating (at 4 ° C during the day, Table 8.1) was also accompanied by the appearance of a characteristic LF fluorescence band with a maximum of 460 nm. This means that the time factor is very important.

The spontaneous process of appearance of LF in MCs significantly depends on temperature. Incubation of MCs for 24 h at 37 ° C gave much more LF than at 4 ° C. An incubation during 2 h at 37 ° C was less effective (Table 8.1). Such a significant temperature dependence means that the total speed of the process is limited not by the lipid peroxidation stage (free radical reactions are not very dependent on temperature), but by some slow stage. It can be assumed that this is the denaturation of proteins and their cross-linking.

Simultaneously with the appearance of a fluorescent peak of LF at 460 nm, a second peak – at 520 nm was detected in a suspension of aging MCs (Fig. 8.2). This peak is due to the release of endogenous flavins (FMN and riboflavin) from MCs into the solution [25] (inside the MC, the flavins fluoresce twice as bad as in the solution [26] due to Tyndale-Rayleigh hypochromism [27]). The release of flavins into the solution is accompanied by an increase in the NADH-induced generation of superoxide, i.e. leads to superoxide-dependent destructive processes in MCs.

When blowing a suspension of MCs with argon, the formation of LF was slower (Table 8.1), because argon displaces oxygen from the solution. This means that the

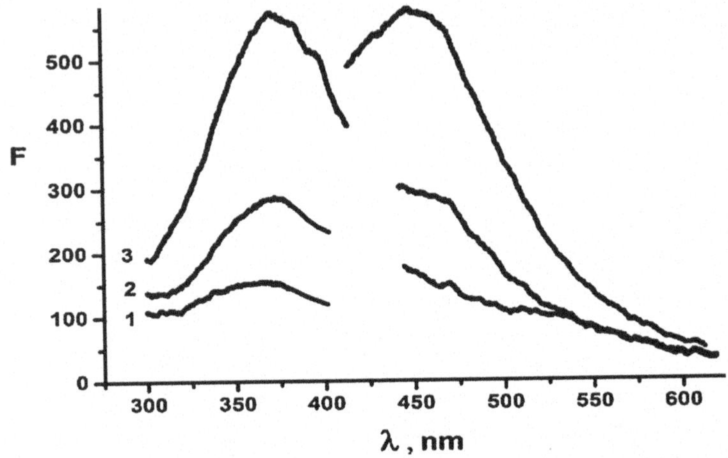

Fig. 8.4 The effect of irradiation on the formation of LF in MCs (0.3 mg/ml): initially (1), 1 h in glass cuvette (2) and 1 h in quartz cuvette (3). Excitation spectra were measured at emission 460 nm; emission spectra were measured at excitation 360 nm

formation of thermo-LF in MCs is oxygen-dependent, similar to lipid peroxidation in the Fenton model reaction [19].

The formation of LF in MCs was sharply activated by powerful light irradiation (Fig. 8.4). Moreover, the greatest contribution was made by UV light in the range 220–320 nm. In a glass cuvette that did not pass UV shorter than 320 nm, the amount of LF formed was much smaller than in a quartz cuvette that was good in UV transmission.

The fluorescence band of LF arising during the 1-h UV irradiation of MCs is characterized by the maximum at 460 nm. The flavin peak at 520 nm was not detected due to photo-degradation of flavins upon strong illumination [18]. The band at 360 nm is visible in the excitation spectrum of LF (Fig. 8.4).

Since the shape of the excitation and emission spectra of photo-LF differs markedly from the shape of the spectra of thermo-LF, this means that these are two different products of degradation of MCs.

When the MC suspension was purged with argon, the amount of photo-LF formed also decreased markedly. This means that the formation of photo-LF occurs with the participation of oxygen. A decrease in the formation of photo-LF was also observed during a preliminary 30-min incubation of MCs with 10 mM succinate in an airtight cuvette (Table 8.1), which is due to oxygen depletion as a result of the MC respiration.

Since the MC aggregation occurred during UV irradiation and during prolonged heating, leading to light scattering, figures for the LF content were obtained under conditions of aggregate destruction by SDS. In [20], the authors worked without detergents. They carried out centrifugation precipitating of the bulk of the particles and then had take for analysis only the supernatant, i.e. they thereby lost most of the LF. SDS destroys not only aggregates, but also membranes, disintegrating protein complexes, but it does not destroy LF, because strong dense chemical bonds formed between the denatured proteins and oxidized lipids products. The addition of SDS sharply reduces light scattering and eliminates Tyndale-Rayleigh hypochromism [26].

In addition to the fluorescence intensity, the polarization degree of the LF fluorescence was measured, the value of which allows determine whether it exists in a bound or unbound state. LF in MC has a rather high polarization degree >0.3, which is due to the low rotational mobility. Polarization degree of LF during 3-h incubation of MCs at 49 ° C did not change, although the amount of LF increased. This means that this LF does not leave the MC membranes into solution (when it enters the water, the degree of polarization should decrease).

A series of control experiments without MCs were also carried out. For this, non-mitochondrial proteins — BSA, trypsin, lysozyme — were used. Their prolonged heating did not lead to the formation of any LF. This means that for the occurrence of thermo-LF, the presence of proteins alone is not enough.

The effect of UV irradiation on non-mitochondrial proteins in solution led to the appearance of characteristic LF-like fluorescence, although it did not cause aggregation. This means that here the photo-LF has an intra-protein nature, and not inter-protein. The fluorescence intensity of the photo-LF is directly proportional to the number of formed intra-protein Schiff bases.

In other control experiments, DNA was tested. When the DNA solution was irradiated for 1 h, the band at 460 nm was very weak (Table 8.1). This means that DNA in MCs is hardly directly involved in the formation of LF.

During UV irradiation of non-mitochondrial proteins (lysozyme, trypsin, BSA), it was proved that the formation of photo-LF does not require the presence of a lipid fraction, just one protein is enough (Table 8.1).

Interestingly, during argon purging, the formation of photo-LF decreased markedly (Table 8.1). Obviously, during UV irradiation, tryptophan and tyrosine protein residues not only give photo-excitation energy to peptide vibrations and conformational changes, but also generate reactive oxygen species that crosslink the protein molecule intramolecularly. Moreover, it seems that tryptophan residues can participate directly – to form a fluorescent chromophore for LF.

Determination of the lipid peroxidation level in the suspension of MCs using the MDA-test showed that 2-h incubation of MCs at 37 ° C leads to a three-fold increase in lipid peroxidation (Table 8.2).

Table 8.2 Lipid peroxidation in a suspension of MCs after heating (it was measured photometrically by a complex of TBA with MDA-like products)

Sample:	D at 532 nm
MCs source	0.1
MCs after 2-h incubation at 37 ° C	0.35
The same in the presence of 0.3 mM succinate	0.24
The same in the presence of 0.3 mM bicarbonate	0.22

Incubation of MCs in the presence of succinate, which activates the MC respiration, or displacement of oxygen using carbon dioxide formed during the spontaneous decomposition of sodium bicarbonate in an aqueous solution at 37 ° C, led to a significant decrease in lipid peroxidation. This correlates with a decrease in the formation of thermo-LF.

Lipid peroxides are necessary for the formation of thermo-LF in MC. That is why thermo-LF is formed in MC, but not in individual proteins. In general, lipids are not needed for the formation of photo-LF.

So, LF can arise in MC spontaneously (without lysosomes and any intracellular substances) due to reactions involving intrinsic iron ions (iron-sulfur clusters and cytochromes), endogenous respiratory chain substrates, lipids and proteins. In living cells, other factors can also be superimposed on this process, weakening or, conversely, accelerating it.

Thermo-lipofuscin and photo-lipofuscin formed in the MC suspension are slightly different from the initial lipofuscin accumulated by aging MCs in the body of adult rats. But many of their properties are very close: (1) they have a blue fluorescence band, (2) they are not a large inter-protein network; (3) judging by the separation on the column with Sephadex, their m.v. less than 70 kDa, i.e. they consist of only 1–2 globules of damaged proteins, (4) they are very resistant to acids and alkalis, to many reducing agents and oxidizing agents, to proteinases, etc. (5) they can be destroyed only by dithionite; (6) they are poorly soluble in water.

8.1.1 Conclusion

It has been shown that isolated MCs can undergo transformation into lipofuscin granules without any additional factors, slowly spontaneously at a low temperature or quickly when heated (thermo-lipofuscin), or under UV irradiation (photo-lipofuscin). The main contribution to the formation of mitochondrial lipofuscin is made by denatured proteins. In this case, the formation of thermo-lipofuscin depends on lipid peroxidation, and the formation of photo-lipofuscin during UV irradiation does not require lipids. The results were obtained in collaboration with M.S. Frolova and published [28].

8.2 Disruption of Flavin Homeostasis in Isolated Mitochondria

A number of studies have demonstrated the spontaneous dissociation of flavin (in the form of FMN) from the 3-subunit flavoprotein fragment of complex I of the respiratory chain [29], from the intact bovine and isolated prokaryotic enzyme [30, 31] and from isolated MCs [32]. A significant change in the angle of the butterfly-shaped structure of the iso-alloxazine ring [33] and the overall weakening of the molecular structure of complex I upon restoration [34] may contribute to a decrease in the affinity of flavin for the binding site.

Dissociation of FMN from complex I in a suspension of isolated mitochondria is observed not only at extreme conditions (lower pH, heating) [35], but also under physiological conditions [32]. The prolonged succinate oxidation by the brain's MCs under conditions of reverse electron flow stimulates the dissociation of flavin from complex I that leads to a loss of enzymatic activity and increases the generation of reactive oxygen species (ROS) [36]. A significant decrease in the level of FMN is observed during retinal degeneration, which leads to death of photoreceptor cells [37]. Presumably, the main reason for the decrease in the activity of complex I after ischemia is the release of flavin from the enzyme [38]. It was noted [39] a marked increase in the concentration of FMN in the blood plasma of critically ill patients. This factor can serve to measure the patient's status [39]. Free reduced flavin $FMNH_2$ can be directly oxidized with oxygen, which leads to an increase in the local concentration of ROS in the matrix.

The release of FMN from complex I is diagnosed by an increase in flavin fluorescence (excitation of 450 nm and emission of 525 nm) [36] and a decrease of its polarized fluorescence [40]. The FMN quantum yield is 0.22, which is 3–5 times more than that of FAD [41], therefore, it is accepted that FMN makes the main contribution to the fluorescent signal from MCs [41]. However, the riboflavin quantum yield is even larger and amounts to 0.27 [41]. Thus, it is obvious that one cannot neglect a possible contribution from riboflavin (RF).

Most of flavoproteins, located on the inner membrane and in matrix, have non-covalently bound FAD, which theoretically could also lose contact with the enzyme [42].

Based on the data of high performance liquid chromatography (HPLC) it was proved that during incubation of isolated rat liver MCs for 2 h, the main process responsible for altering the flavin fluorescence is the hydrolysis of FAD to FMN and riboflavin (RF). The results are presented below.

Free RF and FMN, unlike FAD, could provide non- enzymatic catalyzis of oxidation of NADH by molecular oxygen (dissolved in water) to form a superoxide (even in a dark conditions) [43]. It was also interesting to estimate the amount of non- enzymatic shunting of electrons from NADH through flavin to oxygen.

Rat liver MCs were prepared as described previously [44]. The liver was homogenized in the ice-cold isolation buffer (IB) containing 250 mM sucrose, 20 mM Tris-HCl, 1 mM EGTA (pH 7.7). The sample was then centrifuged at 1000 g for 15 min. The supernatant was centrifuged for 15 min at 5000 g, where mature heavy MCs were precipitated, which were then suspended in the same medium (9 ml per 1.5 ml of pellet) and dosed into Eppendorf tubes. MCs were prepared freshly for each experiment and used within 4 h of isolation, and all steps were strictly operated in ice to guarantee the isolation of high-quality MC preparations. All MC fractions were equilibrated at same protein concentration of 0,3 mg/ml with the use of incubation buffer (100 mM sucrose, 10 mM Tris-phosphate, pH 7.0). Before starting the experiments with NAD and NADH, the MC suspension was subjected to freezing and thawing in isolation buffer. The release of proteins and DNA from MCs into free solution at these conditions was not observed [45]. It means that MC membranes remained generally intact. The concentration of the MC protein was determined by UV-express method [46].

Chemiluminescence measurements were carried out with Lum-5773. Changes of ROS amount in the diluted MC suspension were detected using 2-methyl-6-(4-methoxyphenyl)-3,7- dihydroimidazo [1,2-a]pyrazin-3(7H)-one (MCLA), which is a specific chemiluminescent probe for superoxide [34]. MCLA was added to MCs at 0,3 mg/ml (pH 7.0). A number of substances (NADH, ADP, AMP, GDP, GMP, adenosine, nicotinamide, etc.) were added directly into this MCLA solution with mixing. None of these substances directly affects the chemiluminescence of MCLA either in water or in MCs (data not shown). The kinetics of chemiluminescence was immediately measured at 21° C.

Oxygen consumption was measured using optical oxygen sensor VISIFERM DO 120, HAMILTON. Measurements were performed in medium containing 10 mM Tris-phosphate, pH 7.0.

The LF and flavin fluorescence was carried out with Cary Eclipse.

FAD, FMN and RF were measured in isolated MCs by HPLC with reversed phase and fluorescent detection (excitation and emission = 450 nm and 535 nm, respectively) based on [47]. Because FAD, FMN and RF are light-sensitive, the preparation of sample extracts was carried out in subdued light. 200 ml of MC suspension in selection buffer were mixed with 500 ml of incubation buffer; than 1 ml of mixture of ice-cold methanol: 10 mM K-phosphate (pH 5.0) in ratio of 1:1 was added. This mixture was centrifuged during 4 min at 25000 rpm to separate proteins. Supernatant was diluted and only 20 ml was used for the HPLC run. An eluent mixture contains 10 mM K-phosphate, pH 5.0: methanol in the ratio of 3:1. The retention times were quite reproducible and the recoveries of FAD, FMN and RF were 16.9 ± 0.8, 21.3 ± 0.7, 27.8 ± 1.02, respectively.

8.2.1 Changes in Flavin Fluorescence

During 2-h incubation of MCs at 37 °C, an increase of flavin fluorescence by 2.5 times was observed (Fig. 8.5a) [40]. It is the result not only the release of FMN from complex I, but also due to hydrolysis of FAD in others flavoenzymes. The appearance of RF was detected (Fig. 8.5b). During 2 h time, almost 85% of FAD is hydrolyzed (Fig. 8.5b). Besides, more than half of FMN is hydrolyzed to RF. Since the lifetime and fluorescence quantum yield of FMN and RF are close, this does not make a tangible contribution to changes in the flavin signal in the MC suspension (Fig. 8.5b). RF fluorescence increases by 4 times during the incubation of MCs (Fig. 8.5b).

Interestingly, 1 mM EDTA completely stops the hydrolysis of FAD. The FAD level does not fall during 2 h (Fig. 8.5c). The level of FMN is higher by 63% compared with the control, and the level of RF – by 1.9 times (Fig. 8.5a, c).

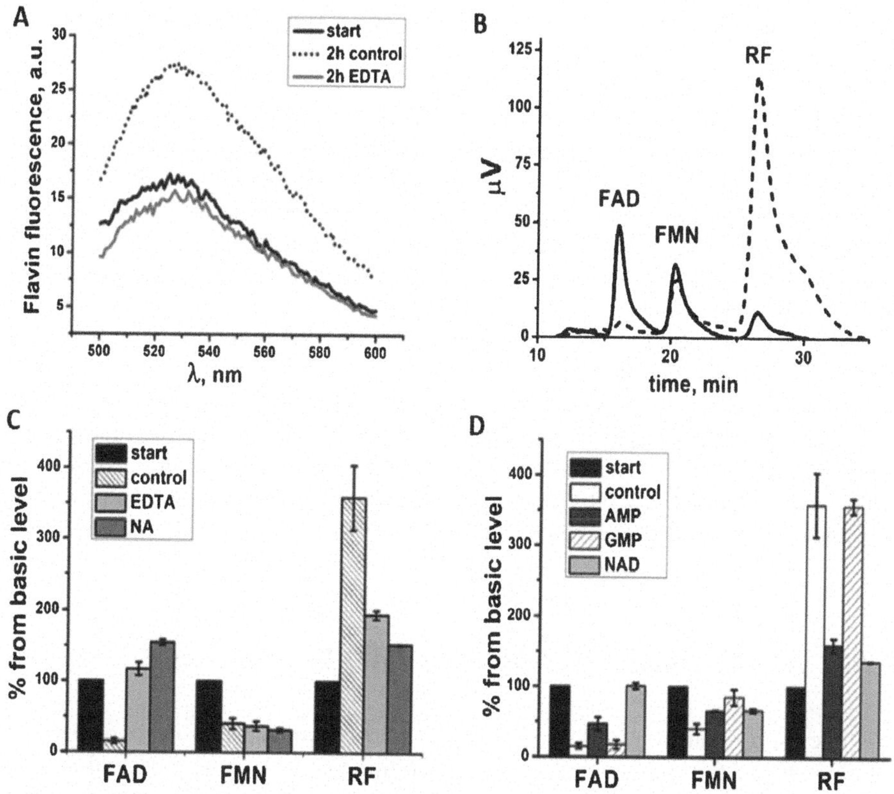

Fig. 8.5 (**a**), Total flavin fluorescence of MCs (excitation 450 nm, emission 525 nm): before the start of incubation (black line), after incubation 2 h at 37 °C (dot line), after incubation 2 h at 37 °C with 1 mM EDTA (grey line). (**b**), HPLC elution profiles of FAD, FMN, and RF after methanol extraction from MCs: before incubation (solid line), after incubation for 2 h at 37 °C (dot line). (**c**), FAD, FMN and RF levels (%) after methanol extraction from MCs after incubation for 2 h at 37 °C: control, without substance addition (striped columns), in the presence of 1 mM EDTA (light gray columns), in the presence of 1 mM nicotinamide (NA) (dark gray columns), start level before incubation (black columns). (**d**), FAD, FMN and RF levels (%) after methanol extraction from MCs after incubation for 2 h at 37 °C: control, without substance addition (white columns), in the presence of 1 mM AMP (dark gray columns), in the presence of 1 mM GMP (striped columns), in the presence of 1 mM NAD (light gray columns), start level before incubation (black columns)

It was noticed that some NAD derivatives, such as AMP, GMP, NAD at a concentration of 100 μM, and nicotinamide at concentration of 1 mM slow down the increase of flavin fluorescence in the MC suspension (Fig. 8.5c, d). Previously, we interpreted this as the effect of these substances on the ability to stabilize only the flavin molecule in complex I. With HPLC data, it became obvious that the whole group of these substances is able to stop the hydrolysis of FAD in many flavoenzymes (Fig. 8.5c, d). Nicotinamide stops hydrolysis of FAD most of all: the concentration of FAD is higher by 55%. The FAD level during incubation with NAD does not change, and with AMP it decreases two times.

GMP preserves FMN by 85% (Fig. 8.5d). This could mean that GMP retains FMN in complex I, which protects it from hydrolysis to RF. EDTA and nicotinamide do not affect the FMN hydrolysis.

It is known that hydrolysis reactions of FAD and FMN in mammalian blood plasma can be accelerated enzymatically using FAD pyrophosphatase and FMN phosphatases, which catalyze the conversion of FAD to FMN and the conversion of FMN to RF, respectively [48, 49]. EDTA inhibits the hydrolysis of FAD and FMN in plasma at 4 °C [49]. Interestingly, the both enzyme activities – FAD pyrophosphatase and FMN phosphatase – were also inhibited by EDTA in solution [50] and MCs (Fig. 8.5c).

Studies with AMP (1 mM), a known inhibitor of FAD pyrophosphatase, gave a similar result (Fig. 8.5d) [51].

Another explanation of the effect of EDTA and AMP on the hydrolysis of flavins is possible: EDTA is a strong chelator of iron ions [52], which may catalyze the hydrolysis of FAD [53]. The addition of AMP and GMP leads to a partial slowdown in the hydrolysis of FAD [40]. It is likely, GMP and AMP, with the help of their phosphate group, also act as iron chelators.

8.2.2 *Non-enzymatic Hydrolysis of FAD by Iron Ions*

It is interesting that iron ions simulate rapid non-enzymatic hydrolysis of FAD in solution (Fig. 8.6a). Immediately upon the addition of 1–10 mM ferric chloride to 100 μM FAD in aqueous solution a slight decrease in flavin fluorescence was detected due to direct quenching by the paramagnetic ion. Then the slow fluorescence increase was occurred. It exceeds the initial level of the flavin signal. The increase was due to the breakdown of FAD to FMN and RF, which have a higher quantum yield and, consequently, more fluorescent signal. The enhance of flavin fluorescence was stopped by 1 mM EDTA, which chelates iron ions. The quenching effect of iron disappeared. In a control, the adding of 1–10 mM ferric chloride to 100 μM FMN leads only to quenching, without the increase after (Fig. 8.6a).

8.2.3 *Non-enzymatic Superoxide Formation by NADH with Flavins*

Free RF, like FMN, unlike FAD, is capable to non- enzymatic (even under dark conditions) catalyzing the oxidation of NADH to form superoxide [32]. In our model experiments using a polarographic hermetic cell, non-enzymatic oxygen consump-

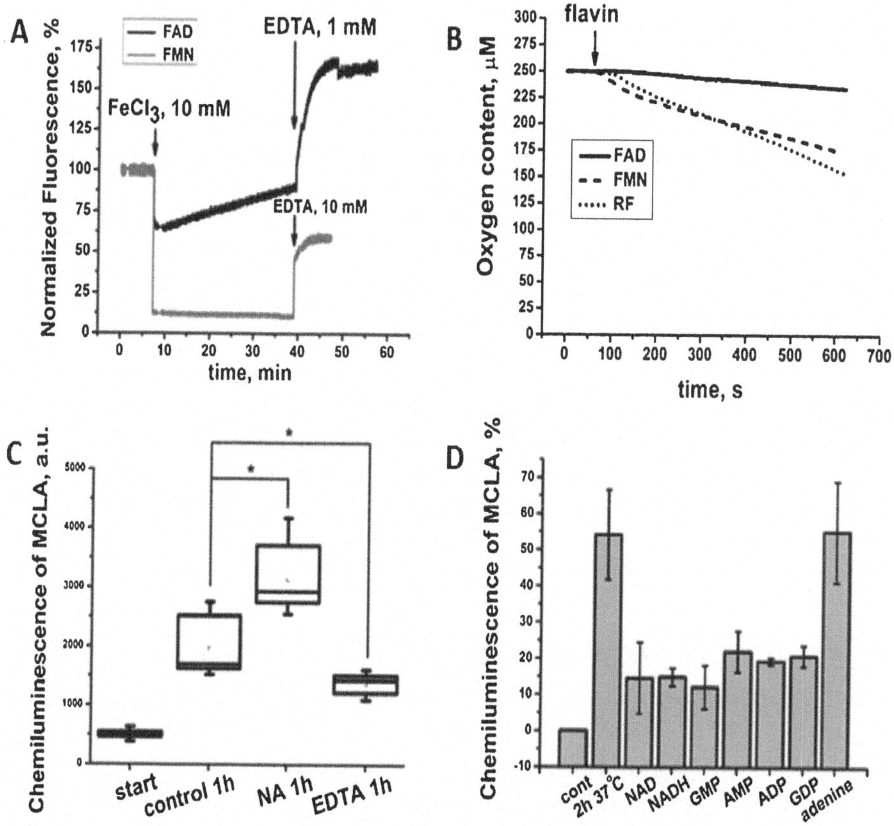

Fig. 8.6 (**a**) Non-enzymatic hydrolysis of FAD by iron ions: aqueous solution of 100 μM FAD (black line), an aqueous solution of 100 μM FMN, control (gray line). (**b**) Non-enzymatic uptake of oxygen by 1 mM NADH solution in 10 mM Tris-phosphate buffer (pH 7.0): solution with addition of 100 μM RF, oxygen uptake is 11.5 mM/min (±0.9) (dash line), solution with addition 100 μM FMN, oxygen absorption is 7.8 μM/min (dot line), solution with addition 100 μM FAD, control (solid line) low consumption for traces of FMN and RF in FAD. (**c**) Effect of incubation of MCs (0.3 mg/ml) for 1 h at 37 °C on chemiluminescence 1 μM MCLA: before incubation (start), after incubation (control), after incubation in the presence of 1 mM nicotinamide (NA) and 1 mM EDTA. (**d**) The effect of pyridine analogs NAD on the formation of superoxide during 2-h incubation at 37°C of MCs (0,3 mg/ml)

tion was observed in a mixture of 1 mM NADH: 100 μM FMN or 1 mM NADH: 100 μM RF in Tris-phosphate buffer (pH 7.0) (Fig. 8.6b). The rate of oxygen consumption in the reaction of RF with NADH is higher than FMN with NADH (Fig. 8.6b). In the course of this reaction, superoxide is formed, which quickly degrades to hydrogen peroxide.

8.2.4 The Effect of Nicotinamide on the FAD Hydrolysis and the Superoxide Formation

An increase of MCLA chemiluminescence is observed during incubation of MCs for 1 h at 37 °C (Fig. 8.6c). The chemiluminescence of MCLA per hour increases by almost 3 times (Fig. 8.6c). Incubation in the presence of 1 mM nicotinamide increases this effect by almost 2 times. Nicotinamide is able to act as a pro-oxidant [54]. Nicotinamide (1 mM) completely stops the hydrolysis of FAD and, it would seem, should have reduced the amount of superoxide, non-enzymatically formed in the reaction of NADH with RF. But here the opposite happens: the concentration of superoxide almost doubles. It means that we cannot completely reject the model of excessive formation of superoxide in the active center of complex I due to the loss of FMN. It is possible that nicotinamide contributes precisely to this process, which does not depend on hydrolysis of FAD. It is known that the nicotinamide part of NAD adversely affects the binding affinity of NADH with the active center of complex I, presumably due to the presence of a positive charge, which has a repulsive force necessary to release the product from the enzyme active center [55].

8.2.5 Effect of EDTA on FAD Hydrolysis and Superoxide Formation

Incubation of MCs during 1 h in the presence of 1 mM EDTA reduces the formation of superoxide (Fig. 8.6c). Like nicotinamide, EDTA completely stops the hydrolysis of FAD. It is known that with neurodegenerative diseases due to heavy metal intoxication, chelation therapy with EDTA administered intravenously is effective [56]. This significantly reduces the level of oxidative stress.

8.2.6 The Effect of Pyridine Nucleotides on the Hydrolysis of FAD and the Superoxide Formation

Incubation of MCs for 2 h at 37 °C with substances containing purine nucleotides (ADP, AMP, GMP, GDP, NADH, NAD) reduces the MCLA chemiluminescent signal by an average of 40% (Fig. 8.6d). Interestingly, hydrophobic adenine, which has not a phosphate group, did not effect on the change in the amount of superoxide (Fig. 8.6d). It would like to remind that these substances in varying degrees affect the hydrolysis of FAD in MCs. NAD completely stops an appearance of superoxide; AMP slows down to almost 50%; GMP retains FAD by only 20%. It can be suggested a considerable contribution of the non-enzymatic reaction with iron ions to the decomposition of FAD and the subsequent non-enzymatic formation of superoxide in the oxidation of NADH by RF. GMP and AMP would work with similar efficiency as iron chelators. Some contribution could make enzymatic hydrolysis of FAD [51].

8.2.7 The Contribution of FMN to the Superoxide Formation

Hydrolysis of FMN to RF makes a minimal contribution to the change of flavin fluorescence, since FMN has a quantum yield similar to RF. In addition, FMN itself is capable to the non-enzymatic production of superoxide during the NADH oxidation (Fig. 8.6b).

GMP (100 μM) significantly reduced superoxide levels during incubation of MCs (Fig. 8.6d). At the same time, FAD is almost not preserved, but 85% FMN is retained (Fig. 8.6d). Thus, here there is a lack of non-enzymatic contribution to the formation of superoxide. Presumably, when FMN remains in the active center of complex I during incubation, as a result, superoxide is not formed by the enzyme. This fact argues in favor of that substances containing adenosine phosphate (ATP, ADP, AMP, NAD and NADH) and guanosine phosphate (GMP, GDP, GMP) prevent the loss of FMN by the active center of complex I [40]. Adenosine and guanosine nucleotides are large molecules of a size which is enough to block an access of oxygen to the reduced flavin. Since they are present in matrix at high local concentrations (the total adenosine nucleotide pool concentration is 10 mM [57]), they compete with NADH for the flavin in complex I binding site and can regulate the rate of superoxide formation [58].

Thus, we observe at least three processes associated with the de-flavinization of flavoproteins and the formation of superoxide in MCs (Fig. 8.7). The first of them takes place in complex I due to FMN release from the active center, resulting in an even greater electron leakage to oxygen. This process is capable of blocking by adenosine phosphates and guanosine phosphates, as well as NAD; but enhances by nicotinamide. The second process is associated with enzymatic hydrolysis of FAD and FMN in the matrix. It is blocked by EDTA, AMP, NA and NAD. The third process is associated with non-enzymatic hydrolysis of FAD by iron/It is blocked by EDTA and AMP. All three processes are closely associated with the formation of ROS.

8.2.8 Conclusion

It has been shown that spontaneous release of non-covalent flavins (from flavoenzymes) begins after isolation of MCs from rat liver. FAD is hydrolyzed to FMN or RF. FMN is hydrolyzed to RF. These processes are stopped by EDTA. In the presence of NADH, de-flavinization of flavoproteins leads to formation of superoxide by at least of three reactions. The first of these occurs in complex I as a

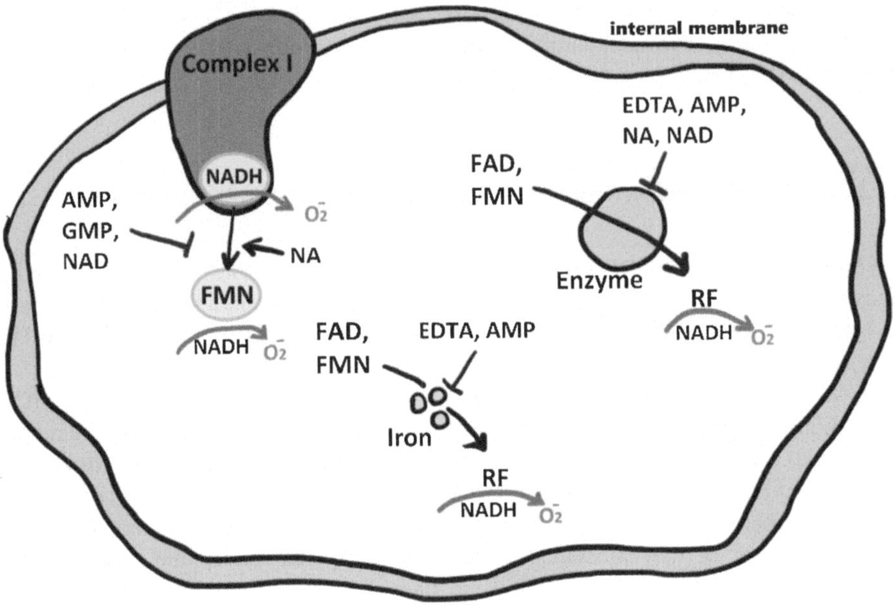

Fig. 8.7 Three processes associated with the de-flavinization of flavoproteins and the formation of superoxide in mitochondria

result of the release of FMN from the active center. This process is inhibited by adenosine phosphates and guanosine phosphates, as well as NAD, but amplified by nicotinamide. The second process is associated with enzymatic hydrolysis of FAD and FMN to RF; it is blocked by EDTA, AMP, NA, NAD. The third process is associated with non-enzymatic hydrolysis of FAD by iron ions in matrix; it is blocked by EDTA and AMP. The results were obtained in collaboration with M.S. Frolova and published in [59].

8.3 Some Properties of Mitochondrial Lipofuscin

Lipofuscin, which accumulates over the years in the MCs of various organs of animals during aging [60], contains little lipids because MCs itself do not contain them too much. Mito-lipofuscin has mainly a protein-source nature [61].

Mito-lipofuscin extracted by isopropanol from MCs (on column G-50 with alcohol) from the liver of an old rat (older 1 year) has a mass not more than 70 kDa (Fig. 8.8).

Mito-lipofuscin is quite hydrophobic: it is poorly soluble in water, but not too bad soluble in detergent aqueous solutions, as well as in alcohols.

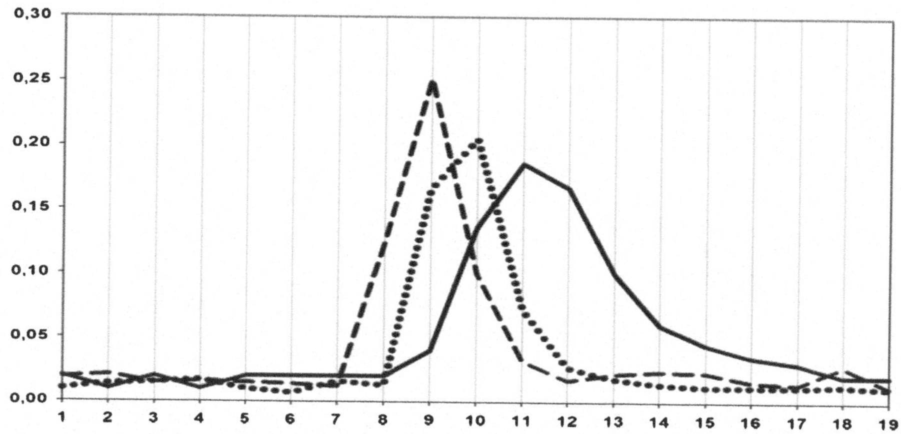

Fig. 8.8 The distribution of protein and lipofuscin in fractions (per 1 ml) obtained from the G-50 column (with 50% alcohol) in the presence of SDS. The MC proteins were detected by optical density at 286 nm (- - -); mito-lipofuscin was detected by fluorescence upon excitation of 360 and emission of 460 nm (—); control: BSA (69 kDa), optical density detection was at 286 nm (. . .)

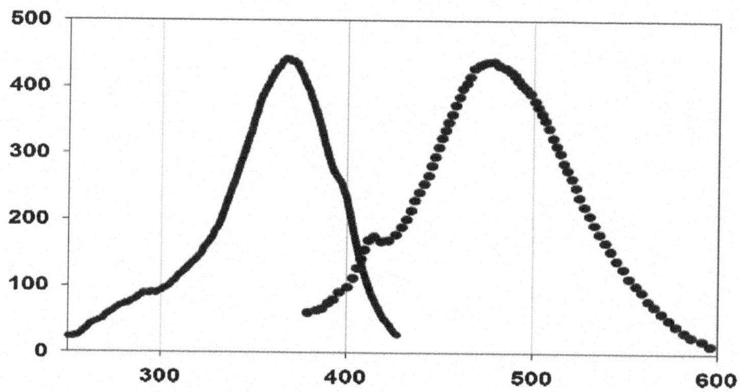

Fig. 8.9 Spectra (in the region of 250–600 nm) of excitation (–) and emission (•••) of mito-lipofuscin

Mito-lipofuscin consists of a chain of damaged fragments of proteins cross-linked like Schiff bases, giving a blue fluorescence band upon excitation at near UV (Fig. 8.9). The maximum of the excitation spectrum is at 365 nm, and the emission is at 470 nm, which is close to the spectra of mito-lipofuscin in whole MCs.

It cannot be excluded that anthranilic acid and poly-Trp formed during the degradation of tryptophan residues also contribute to the blue fluorescence of mito-lipofuscin.

Infrared spectrum of mito-lipofuscin is very similar to IR spectrum of poly-Trp. (Fig. 8.10). The duble peak at 2900 cm^{-1} belongs to CH-vibrations. It can be

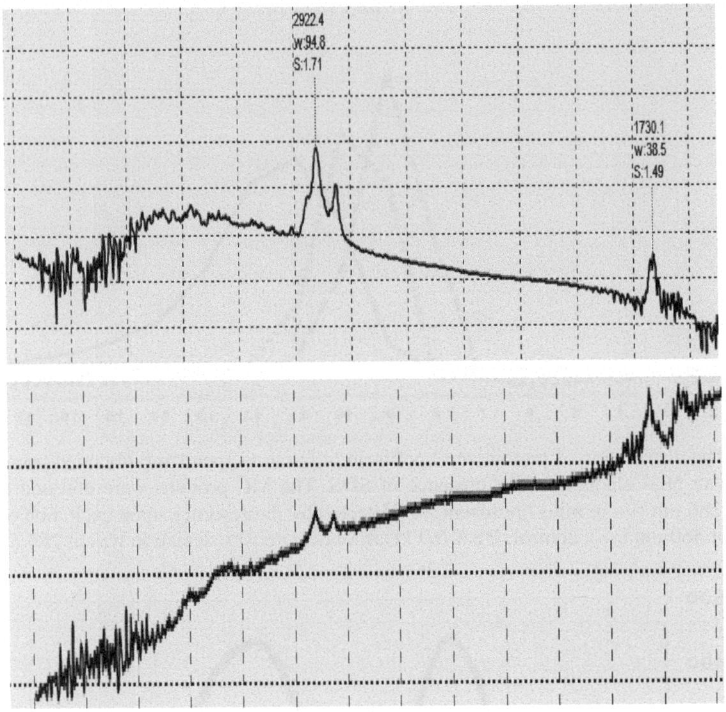

Fig. 8.10 IR absorption spectra of mito-lipofuscin (above) and poly-Trp (below). The dry samples were placed on the CaF_2 window- plate into spectrometer FT-801. The difference in backgrounds of spectra is due to different light-scattering of the samples

suggested that mito-LF is an analog of melanin (the aging pigment, appeared in old cells due to polymerization of aromatic amino acids).

Mito-lipofuscin is very resistant to acids and alkalis: at pH 1 and pH 12 it was not destroyed.

It is resistant to many reducing agents and oxidizing agents. It was not destroyed by millimolar concentrations of ascorbic acid, sodium succinate, NADH and glutathione.

It is resistant to proteolytic enzymes. Neither proteinase nor acid phosphatase nor alkaline phosphatase nor binase affected it. That is why mito-lipofuscin accumulates in cells during aging of animals and is not removed by anything.

It was possible to destroy mito-lipofuscin only by dithionite, a powerful reducing agent. It can be assumed that strong bonds between protein globules in mito-lipofuscin are formed with the participation of sulfhydryl groups (and may be also by tryptophan and tyrosine).

Partial destruction of mito-lipofuscin was carried out using UV irradiation. Figure 8.11 shows the photo-bleaching of mito-lipofuscin under the influence of UV light 360 nm from a 450-watt xenon lamp. The destruction of mito-lipofuscin by UV light was irreversible: even half an hour after the completion of photo-bleaching, the fluorescence intensity did not return to the original one.

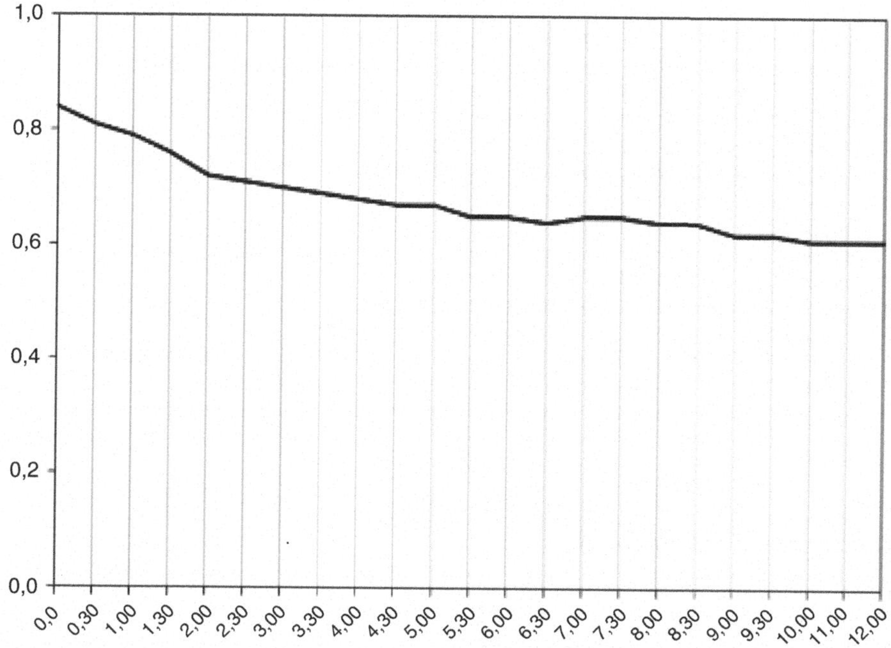

Fig. 8.11 Kinetics (during 12 min) of photo-bleaching of mito-lipofuscin (in water with 0.5% SDS). It was measured in mirror micro-cuvette

The results were obtained in collaboration with A.V. Chaplygina.

In the next future, the properties of mito-lipofuscin will be studied more detail.

References

References Section for 8.1

1. Höhn A, Grune T (2013) Redox Biol 1:140–144
2. Brunk UT, Terman A (2002) Eur J Biochem FEBS 269:1996–2002
3. Brahn B, Schmidtmann M (1920) Pigmentstudien, Virchows Arch. In, CCXXVII
4. Hueck WL, Krehl F, Marchand B (1921) Hndb Allgem Pathologie, Hrsg., III
5. Schutt F et al (2002) FEBS Lett 528(1–3):217–221
6. Warburton S, Southwick K et al (2005) Mol Vis 11:1122–1134
7. Azarian SM et al (2006) J Proteome Res 5:521–529
8. Fishkin NE et al (2005) Proc Natl Acad Sci USA 102(20):7091–7096
9. Tsai L, Szweda PA, Vinogradova O, Szweda LI (1998) PNAS 95(14):7975–7980
10. Jung T et al (2007) Ann N Y Acad Sci 1119:97–111
11. Terman A, Brunk UT (1998) APMIS 106(2):265–276
12. Hohn A, Sittig A, Jung T, Grimm S, Grune T (2002) Free Radic Biol Med 53:1760–1769

13. Höhn A, Jung T, Grimm S, Grune T (2010) Free Radic Biol Med 48:1100–1108
14. Popov GA, Konev VV (1979) Biophysics 24:168–170
15. von Zglinicki T, Nillson E, Docke WD, Brunck UT (1995) Gerontology 41(2):95–108
16. Sapezhinsky II (1979) Biophysics 24:386–391
17. Watanabe Y et al (1991) Chem Pharm Bull 39:1796–1801
18. Dillard CJ, Tappel AL (1971) Lipids 6:715–721
19. Sharova IV, Vekshin NL (2004) Biophysics 49(5):814–821
20. Chio KS et al (1974) Curr Res 2:182–184
21. Vekshin NL (1988) Biol Sci 4:107–111
22. Vekshin NL (2009) Fluorescence spectroscopy of biopolymers. Photon, Pushchino
23. Vekshin NL et al (2007) Meat Technol 3:44–45
24. Sokolova IB, Vekshin NL (2008) Biophysics 53(1):73–77
25. Frolova MS, Vekshin NL (2014) J Fluoresc 24(4):1061–1066
26. Vekshin NL, Frolova MS, Kovalev VI, Begunova EA (2015) Biophysics 60(1):129–135
27. Vekshin NL (2002) Photonics of biopolymers. Springer, Berlin/Heidelberg
28. Frolova MS, Surin AM, Braslavsky AV, Vekshin NL (2015) Biophysics 60(5):1125–1131

References Section for 8.2

29. Sled VD, Vinogradov AD (1993) BBA 1143(2):199e203
30. Gostimskaya IS et al (2007) FEBS Lett 581(30)
31. Holt PJ et al (2016) BBA-Bioenergetics 1857(11)
32. Vekshin NL (2009) Fluorescence spectroscopy of biomacromolecules. Pushchino, Photon
33. Lennon BW, Williams CH, Ludwig ML (2008) Protein Sci 8(11)
34. Mamedova AA et al (2004) J Biol Chem 279(22)
35. Rao NA, Felton SP (1963) Flavin Mononucleotide 238:1
36. Stepanova A (2017) et al Potential mechanism for brain ischemia reperfusion injury
37. Kelley RA et al (2017) J Biol Chem 292(51)
38. Kahl A et al (2018) Stroke 49(5)
39. Vasilaki AT et al (2010) Clin Chim Acta 411(21–22):1611–1624
40. Frolova MS, Vekshin NL (2014) J Fluoresc 24(4)
41. Galba'n J et al (2016) Method Appl Fluoresc IOP Pub 4(4)
42. Lienhart W-D, Gudipati V, Macheroux P (2013) Arch Biochem Biophys 535(2)
43. Vekshin NL (2002) Photonics of biopolymers. Springer, Berlin
44. Sharova IV, Vekshin NL (2004) Biophysics 9(5):814–821
45. Frolova MS, Vekshin NL (2014) Biochem Suppl Ser A Membr Cell Biol 8(2)
46. Redmile-Gordon MA (2013) et al Soil Biol Biochem 67(100). Elsevier
47. Akimoto M et al (2007) J Heal Sci 53(3):332–338
48. Okumura M, Yagi K (1980) J Nutr Sci Vitaminol 26(3)
49. Akimoto M et al (2006) Biol Pharm Bull 29(8)
50. Daniel H, Binninger E, Rehner G (1983) Int J Vitam Nutr Res 53(1)
51. Barile M et al (1997) Eur J Biochem 249(3)
52. Aboulroos SA, El Beissary EA, El Falaky AA (1983) Agro-Ecosystems 8(3e4)
53. Lomtev AS, Bobrov AG, Vekshin NL (2004) Russ J Bioorg Chem 30(3)
54. Higashi-Okai K et al (2006) J UOEH 28(4)
55. Berrisford JM, Sazanov LA (2009) J Biol Chem 284(43)
56. Fulgenzi A et al (2014) Biomed Res Int 2014:281510
57. Schild L et al (1999) Biochim Biophys Acta 1413(1)
58. Birrell JA, Hirst J (2013) Biochem Am Chem Soc 52(23)
59. Frolova M, Marchenkov V, Vekshin N (2019) Biochem Biophys Res Commun 516:1211–1215

References Section for 8.3

60. Chaplygina AV, Vekshin NL (2019) Biophysics 64(1):83–88
61. Frolova MS, Surin AM, Braslavsky AV, Vekshin NL (2015) Biophysics 60(5):1125–1131

The manufacturer's authorised representative in the EU is Springer Nature Customer Service Centre GmbH, Europaplatz 3, 69115 Heidelberg, Germany. If you have any concerns regarding our products, please contact ProductSafety@springernature.com

Printed and bound by CPI Group (UK) Ltd, Croydon, CR0 4YY
22/09/2026
02221058-0008